T0189381

Springer Theses

Recognizing Outstanding Ph.D. Research

Aims and Scope

The series "Springer Theses" brings together a selection of the very best Ph.D. theses from around the world and across the physical sciences. Nominated and endorsed by two recognized specialists, each published volume has been selected for its scientific excellence and the high impact of its contents for the pertinent field of research. For greater accessibility to non-specialists, the published versions include an extended introduction, as well as a foreword by the student's supervisor explaining the special relevance of the work for the field. As a whole, the series will provide a valuable resource both for newcomers to the research fields described, and for other scientists seeking detailed background information on special questions. Finally, it provides an accredited documentation of the valuable contributions made by today's younger generation of scientists.

Theses are accepted into the series by invited nomination only and must fulfill all of the following criteria

- They must be written in good English.
- The topic should fall within the confines of Chemistry, Physics, Earth Sciences, Engineering and related interdisciplinary fields such as Materials, Nanoscience, Chemical Engineering, Complex Systems and Biophysics.
- The work reported in the thesis must represent a significant scientific advance.
- If the thesis includes previously published material, permission to reproduce this must be gained from the respective copyright holder.
- They must have been examined and passed during the 12 months prior to nomination.
- Each thesis should include a foreword by the supervisor outlining the significance of its content.
- The theses should have a clearly defined structure including an introduction accessible to scientists not expert in that particular field.

More information about this series at http://www.springer.com/series/8790

Julian Barreiro-Gomez

The Role of Population Games in the Design of Optimization-Based Controllers

A Large-scale Insight

Doctoral Thesis accepted by
Universitat Politècnica de Catalunya, Barcelona, Spain and
Universidad de Los Andes, Bogotá, Colombia

 Springer

Author
Dr. Julian Barreiro-Gomez
Learning and Game Theory Laboratory,
 Engineering Division
New York University
Abu Dhabi, United Arab Emirates

and

Department of Automatic Control,
 Institut de Robòtica i Informàtica
 Industrial (CSIC-UPC)
Universitat Politècnica de Catalunya
Barcelona, Spain

and

Departamento de Ingeniería Eléctrica y
 Electrónica
Universidad de los Andes
Bogotá, Colombia

Supervisors
Prof. Dr. Carlos Ocampo-Martinez
Department of Automatic Control, Institut de
 Robòtica i Informàtica Industrial
 (CSIC-UPC)
Universitat Politècnica de Catalunya
Barcelona, Spain

Prof. Dr. Nicanor Quijano
Departamento de Ingeniería Eléctrica y
 Electrónica
Universidad de los Andes
Bogotá, Colombia

ISSN 2190-5053   ISSN 2190-5061   (electronic)
Springer Theses
ISBN 978-3-030-06384-9   ISBN 978-3-319-92204-1   (eBook)
https://doi.org/10.1007/978-3-319-92204-1

This Springer imprint is published by the registered company Springer Nature Switzerland AG
The registered company address is: Gewerbestrasse 11, 6330 Cham, Switzerland

Firstly to God,
to my parents,
to my brother,
to my girlfriend.

Supervisors' Foreword

Challenges in modern control engineering have motivated researchers around the world to seek newer perspectives that allow improving the performance of dynamical systems in diverse strategic areas such as communications, transport, water, and energy. Traditional control methods have demonstrated their effectiveness toward achieving that end while taking into account aspects related to stability, reliability, robustness, safety, among others. Many strategies have been reported in the literature for integrating the proper handling of nonlinear dynamics, constraints, high-dimensional models, multi-objective control problems, and other potentially complex features of complex dynamical systems. Among these strategies, optimization-based control methods have reached an outstanding position in the rank of preferred industrial control techniques. In particular, model predictive control has gained the attention not only of the industry but also of diverse fields given the important advances in technology related to calculation machines and other devices devoted to solve optimization problems behind the controller's design. In order to overcome potential computational issues that would be found while solving these optimization problems for uncertain dynamical systems affected by exogenous disturbances, smart methods that allow stating such problems in a clever and convenient way should be considered in order to avoid either increasing the resultant computational burden or getting infeasible solutions, these latters causing the malfunctioning of the entire closed-loop setup.

One area that bears a close relationship with optimization-based techniques is game theory. Game theory studies interactions between self-interested agents and tackles the problem of interaction among agents using different strategies who wish to maximize their welfare. There are different types of games, and among these there is evolutionary game theory (EGT). EGT was originally developed in the 1970s by the seminal work of Maynard-Smith and Price, where they were able to relate the ideas that were developed in economy and the altruistic behavior that emerges in nature. Since then, different contributions have appeared in biology, but in the last years there have been several developments in economy. From this economic perspective, EGT is a dynamic process that describes the behavior of agents in time that combines population games (which describe strategic interactions among a large number of small and anonymous agents) and a revision protocol (which specifies how agents can choose and change strategies). The dynamics that are captured by this process are suitable to model strategic interactions of agents in large-scale systems. From this perspective, the use of EGT in engineering problems has several advantages. In one hand, there is a close relation between the available control inputs with strategies, and the control objective could be expressed in terms of payoff functions. On the other hand, there is a close relation between the solution of a game given by a Nash equilibrium and optimization problems. In this regard, it has been proven that, under certain conditions, the Nash equilibrium satisfies the Karush–Kuhn–Tucker (KKT) first-order conditions of constrained optimization problems. This property has been exploited in order to deal with distributed optimization problems, using potential games and population dynamics. And finally, the solution of games can be obtained by employing just local information, i.e., local rules are designed to achieve a global objective.

Motivated by these ideas, this thesis appeared as the excuse to demonstrate how this couple of strong fields (optimization-based control and evolutionary game theory) can dance together in a great harmony and synchronization interacting among them to obtain controller designs able to further improve the closed-loop performance of dynamical engineering systems, in particular, of large-scale nature. Relevant issues and novel control structures such as non-centralized schemes (distributed and hierarchical decentralized) are proposed, discussed, and implemented in simulation, applied to process control systems of water networks.

As one of the central contributions of this thesis, it can be highlighted the work devoted to cope with the design of distributed control schemes for large-scale systems considering their time-varying partitioning of the monolithic models into smaller sub-systems. This partitioning procedure, also based on evolutionary game theory, allows reducing the complexity of the local controllers and their computational burden, while managing the whole system in a modular way toward increasing the inherent robustness of the entire scheme facing uncertainties, disturbances, and/or faults.

All the dedication invested in developing the research topics proposed in this thesis saw its successful results in more than 20 papers in high-quality journals papers and prestigious conferences/congresses worldwide. This fruitful scientific production performed in just three years demonstrates the amount of ideas behind the thesis and its quality and appealing within the scientific community. Hopefully, in the future, there will be several contributions that will take this thesis as their inspiration to develop more real-time control strategies for large-scale systems.

Barcelona, Spain   Prof. Dr. Carlos Ocampo-Martinez
Bogotá, Colombia   Prof. Dr. Nicanor Quijano
May 2018

Foreword from CEA, the Automatic Control Spanish Association

It is a great pleasure for us to introduce here the work of Julian Barreiro-Gomez, awarded with the best Ph.D. Thesis in Control Engineering during the 2017 edition of the award call organized by the control engineering group of Comité Español de Automática (CEA), the Spanish Committee of Automatic Control.

This yearly award is aimed at recognizing outstanding Ph.D. research carried out within the Control Engineering field. At least one of the supervisors must be a partner of the Spanish Committee of Automatic Control and member of the Control Engineering group. The jury is composed of three well-known doctors in the control engineering field; two of them are partners of the CEA, while the third one is a foreign professor.

The first edition counted submissions from Ph.D. students examined between July 2016 and July 2017. Notably, the scientific production of the candidates consisted of a total of 21 publications in international indexed journals, where 19 of them are ranked in first quartile journals. This shows the high scientific quality of the submitted PhDs. All the thesis presented:

- have been under the international doctorate mention (the candidate must have at least one international research stay out of Spain, write part of the thesis in English, and have foreign members in the Ph.D. jury)
- have been mostly oriented to academic research than to industrial applications
- were presented at international conferences. Among them, IFAC and IEEE are considered as preferred conferences.

As a result, Julian Barreiro-Gomez's꼽 Ph.D. thesis was selected as the best one among an excellent group of candidates. His Ph.D. thesis deserves the label "the best of the best," Springer Theses' main motto. Not only his nomination as winner of the Spanish control engineering context has been our great pleasure, but also to recommend his thesis for publication within the Springer Theses collection. On behalf of CEA, we wish Julian to continue his outstanding scientific career, keeping his genuine enthusiasm. We also hope that his work may be of inspiration for other students working in the control engineering field.

Barcelona, Spain

Prof. Dr. Joseba Quevedo
President of CEA
Dr. Ramon Vilanova
Coordinator of the Control
Engineering Group, Comité Español
de Automática (CEA)
Dr. Jose Luis Guzman
Coordinator of the Control
Engineering Group, Comité Español
de Automática (CEA)

Some parts of this thesis have been published in the following articles:

Refereed Journal Papers

1. J. Barreiro-Gomez, G. Obando and N. Quijano. Distributed population dynamics: Optimization and control applications. *IEEE Transactions on Systems, Man, and Cybernetics: Systems* 47(2): 304–314, 2017.
2. N. Quijano, C. Ocampo-Martinez, J. Barreiro-Gomez, G. Obando, A. Pantoja, and E. Mojica-Nava. The role of population games and evolutionary dynamics in distributed control systems. *IEEE Control Systems Magazine*, 37(1): 70–97, 2017.
3. J. Barreiro-Gomez, C. Ocampo-Martinez, and N. Quijano. Dynamical Tuning for MPC Using Population Games: A Water Supply Network Application. *ISA Transactions*, 69(2017): 175–186, 2017.
4. J. Barreiro-Gomez, C. Ocampo-Martinez, N. Quijano and J. M. Maestre. Non-centralized Control for Flow-based Distribution Networks: A Game-theoretical Insight. *Journal of the Franklin Institute*, 354(14): 5771–5796, 2017.
5. J. Barreiro-Gomez, N. Quijano and C. Ocampo-Martinez. Constrained distributed optimization: A population dynamics approach. *Automatica*, 69(2016): 101–116, 2016.
6. G. Riaño-Briceño, J. Barreiro-Gomez, A. Ramirez-Jaime, N. Quijano and C. Ocampo-Martinez. MatSWMM - An open-source toolbox for designing real-time control of urban drainage systems. *Environmental Modelling & Software*, 83: 143–154, 2016.

Refereed Conference Papers

1. J. Barreiro-Gomez, C. Ocampo-Martinez, and N. Quijano. Partitioning for Large-scale Systems: A Sequential DMPC Design. *20th IFAC World Congress*, Vol 50 of IFAC Papers Online, pp. 8838–8843, Elsevier 2017.
2. J. Barreiro-Gomez, C. Ocampo-Martinez, and N. Quijano. On the Communication Discussion of Two Distributed Population-game Approaches for Optimization Purposes. *20th IFAC World Congress*, Vol 50 of IFAC Papers Online, pp. 11782–11787, Elsevier 2017
3. J. Barreiro-Gomez, I. Mas, C. Ocampo-Martinez, R. Sánchez Peña and N. Quijano. Distributed formation control of multiple unmanned aerial vehicles over time-varying graphs using population games, *In Proceedings of the 55th IEEE Conference on Decision and Control (CDC)*, 2016, Las Vegas, USA, pp. 5245–5250.
4. J. Barreiro-Gomez, N. Quijano and C. Ocampo-Martinez. Distributed MPC with time-varying communication network: A density-dependent population games approach, *In Proceedings of the IEEE 55th Conference on Decision and Control (CDC)*, 2016, Las Vegas, USA, pp. 6068–6073.

5. G. Obando, J. Barreiro-Gomez and N. Quijano. A class of population dynamics for reaching epsilon-equilibria: Engineering applications, *In Proceedings of the American Control Conference (ACC)*, 2016, Boston, USA, pp. 4713–4718.

6. J. Barreiro-Gomez, C. Ocampo-Martinez, J. M. Maestre and N. Quijano. Multi-objective model-free control based on population dynamics and cooperative games, *In Proceedings of the 54th IEEE Conference on Decision and Control (CDC)*, 2015, Osaka, Japan, pp. 5296–5301.

7. J. Barreiro-Gomez, G. Obando, C. Ocampo-Martinez and N. Quijano. Making non-centralized a model predictive control scheme by using distributed Smith dynamics, *In Proceeding of the 5th IFAC Conference on Nonlinear Model Predictive Control (NMPC)*, 2015, Seville, Spain, Vol 48:23 of IFAC-PapersOnLine, pp. 501–506, 2015.

8. J. Barreiro-Gomez, G. Obando, G. Riaño-Briceño, N. Quijano and C. Ocampo-Martinez. Decentralized control for urban drainage systems via population dynamics: Bogota case study, *In Proceedings of the European Control Conference (ECC)*, 2015, Linz, Austria, pp. 2426–2431.

9. J. Barreiro-Gomez, N. Quijano and C. Ocampo-Martinez. Distributed resource management by using population dynamics: Wastewater treatment application, *In Proceedings of the 2nd IEEE Colombian Conference on Automatic Control*, 2015, Manizales, Colombia, pp. 1–6.

Refereed Book Chapter

1. J. Barreiro-Gomez, G. Riaño-Briceño, C. Ocampo-Martinez and N. Quijano. Data-driven Evolutionary-game-based Control for Drinking Water Networks. *Real-time Monitoring and Operational Control of Drinking Water Systems.* Springer, 2017.

2. J. Barreiro-Gomez, C. Ocampo-Martinez and N. Quijano. Evolutionary-game-based dynamical tuning for multi-objective model predictive control. In S. Olaru, A. Grancharova, F. Lobo Pereira (editors), *in Model-Based Optimization and Control*, chapter 6. Springer Verlag, 2016.

Acknowledgements

It is time to finish this doctoral journey. It has been a period of my life full of anecdotes and experiences. While doing research, I interacted with many people from whom I learned many academic things and with others from whom I always received support and encouragement. I do want to express my gratitude to those who participated in this journey.

First, I want to thank my advisors and friends Prof. Nicanor Quijano from Universidad de los Andes (UniAndes) and Prof. Carlos Ocampo-Martinez from Universitat Politècnica de Catalunya (UPC) for their support during this journey. They have contributed to my professional training with their critiques and persistent questioning, being always available for academic discussions. They have also given me the possibility to begin this challenging and wonderful journey from which I have learned plenty of things.

Second, I want to thank Prof. Jason Marden from University of California at Santa Barbara (UCSB) for receiving me as a visitor researcher and for giving me a place to work within the Control and Computation Research Laboratory and the opportunity to interact with excellent researchers. I thank Prof. Philip Brown for sharing novel research ideas with me. Special thanks go to Prof. Jorge Poveda for investing time entering into details in academic discussions from which I learned a lot and for suggesting formal alternatives to address new research challenges. I also want to thank Prof. Ricardo Sanchez Peña and Dr. Ignacio Mas for having invited me to their research group at Instituto Tecnológico de Buenos Aires (ITBA) in order to work on formation control for unmanned aerial vehicles. I want to thank Prof. Germán Obando from Pontificia Universidad Javeriana for all the discussions about our research trying to find new ideas while he was a Ph.D. student at UniAndes. I would like to thank Dr. Fernando Bianchi from Instituto de Investigación en Energía de Cataluña (IREC) and Prof. José María Maestre from Universidad de Sevilla since we performed some specific research.

Third, I thank Prof. Bart De Schutter from Delft University of Technology, and Prof. Sorin Olaru from Centrale Supélec for having reviewed this thesis providing valuable comments to improve the quality of this manuscript. Moreover, I would like to thank Prof. Robert Griñó from UPC and Prof. Alain Gauthier from UniAndes for joining my thesis committee.

Fourth, I sincerely thank Agència de Gestió d'Ajust Universitaris i de Recerca AGAUR for having granted the FI scholarship for me to study the doctoral degree at UPC, and Colciencias for having granted the scholarship/condonable loan (Ref. 6172) for me to study the doctoral degree at UniAndes. I would also like to thank the projects ECOCIS: EConomic Operation of Critical Infrastructure Systems (Ref. DPI2013-48243-C2-1-R), EFFINET: EFFicient Integrated real-time monitoring and control of drinking water NETworks (Ref. FP7-ICT-2011-8-318556), ModCon: Modelado y control de redes de alcantarillado: aplicación en Bogotá (Colombia), and "Drenaje urbano y cambio climático: hacia los sistemas de alcantarillado del futuro. Fase II," Mexichem, Colombia, for partially supporting me.

Finally, *but being the most important acknowledgment*, I owe my gratitude to my parents, Luis Alberto Barreiro Lesmes and Margarita María Gómez Díaz, for having taught me the really important things in life. Essential aspects that are not learned in the academic environment, but that without them, the successful culmination of this journey would not have been possible. I do thank them for having taught me to have perseverance and to pursue goals in an honest manner. I also thank my parents for their words in tough moments throughout this journey, for listening to me and for giving wise advice to me. Likewise, I would like to thank my brother, Juan Esteban Barreiro Gómez, for being exemplar of discipline. I also express my thankfulness to Mayerly Espitia Espinosa, who has always supported and encouraged me since we met. She has always made me smile in stressful moments and has provided a lot of happiness to my days with her love.

Barcelona, Spain   Dr. Julian Barreiro-Gomez
2017

Contents

Acronyms

BWSN	Barcelona Water Supply Network
CMPC	Centralized Model Predictive Control
CSP	Constrained Satisfaction Problem
CSTR	Continuous Stirred Tank Reactor
D3PD	Distributed Density-dependent Projection Dynamics
D3RD	Distributed Density-dependent Replicator Dynamics
D3SD	Distributed Density-dependent Smith Dynamics
DDPG	Density-dependent Population Games
DLD	Distributed Logit Dynamics
DMPC	Distributed Model Predictive Control
DPD	Distributed Projection Dynamics
DRD	Distributed Replicator Dynamics
DSD	Distributed Smith Dynamics
KPI	Key Performance Index
LMPC	Local Model Predictive Control
MPC	Model Predictive Control
QP	Quadratic Programming
UAV	Unmanned Aerial Vehicle

Notation

$\{\cdots\}$	Set
$\emptyset$	Empty set
$i \in \mathcal{X}$	i is an element of the set $\mathcal{X}$
$\mathbb{R}$	Set of real numbers
$\mathbb{R}_{>c}$	$\mathbb{R}_{>c} \triangleq \{x \in \mathbb{R} : x > c\}$ for some $c \in \mathbb{R}$
$\mathbb{R}_{\geq c}$	$\mathbb{R}_{\geq c} \triangleq \{x \in \mathbb{R} : x \geq c\}$ for some $c \in \mathbb{R}$
$\mathbb{R}^n$	Space of n-dimensional (column) vectors with real entries
$\mathbb{R}^{n \times m}$	Space of n by m matrices with real entries
$\mathbb{Z}$	Set of integers
$\mathbb{Z}_{>c}$	$\mathbb{Z}_{>c} \triangleq \{x \in \mathbb{Z} : x > c\}$ for some $c \in \mathbb{Z}$
$\mathbb{Z}_{\geq c}$	$\mathbb{Z}_{\geq c} \triangleq \{x \in \mathbb{Z} : x \geq c\}$ for some $c \in \mathbb{Z}$
$\mathcal{X}(\subset) \subseteq \mathcal{Y}$	Set $\mathcal{X}$ is a (strict) subset of $\mathcal{Y}$
$\mathcal{X} \times \mathcal{Y}$	Cartesian product of the sets $\mathcal{X}$ and $\mathcal{Y}$, i.e., $\mathcal{X} \times \mathcal{Y} = \{(x,y) : x \in \mathcal{X}, y \in \mathcal{Y}\}$
$\mathcal{X}^n$	n-dimensional Cartesian product $\mathcal{X} \times \mathcal{X} \times \ldots \times \mathcal{X}$, for some $n \in \mathbb{Z}_{\geq 1}$
$\text{int}(\mathcal{X})$	Interior of the set $\mathcal{X}$
$\mathrm{x}^\top (\mathbf{X}^\top)$	Transpose of a vector $\mathrm{x} \in \mathbb{R}^n$ (matrix $\mathbf{X} \in \mathbb{R}^{n \times m}$)
$\mathbf{X}^{-1}$	Inverse of the matrix $\mathbf{X} \in \mathbb{R}^{n \times n}$
x_i	i-th element of the vector $\mathrm{x} \in \mathbb{R}^n$
$[a_{ij}]$	Element in the i-th row and j-th column of the matrix $\mathbf{A} \in \mathbb{R}^{n \times m}$
$\text{diag}(\cdot)$	Operator that builds a diagonal matrix with the elements of its argument
$n!$	Factorial of n, where $n \in \mathbb{Z}_{>0}$
$|\cdot|$	(1) $|x| \geq 0$ with $x \in \mathbb{R}$ returns the absolute value of the scalar x
	(2) $|\mathcal{A}|$ returns the cardinality of the set $\mathcal{A}$
$\|\cdot\|$	2-norm (Euclidian norm) of a vector, i.e., $\|x\| \triangleq \sqrt{\sum_{i=1}^{n} x_i^2}$, where $x \in \mathbb{R}^n$

$\mathbb{I}_n$	Identity matrix of dimension $n \times n$, where $n \in \mathbb{Z}_{\geq 1}$
$\mathbb{1}_n$	Vector with $n \in \mathbb{Z}_{\geq 1}$ unitary entries, i.e., $\mathbb{1}_n = \begin{bmatrix} 1 & \cdots & 1 \end{bmatrix}^\top \in \mathbb{R}^n$
$0_{m \times n}$	Zero matrix of dimension $m \times n$, where $m, n \in \mathbb{Z}_{\geq 1}$
$\mathbf{X} \succ 0 \ (\prec 0)$	$\mathbf{X}$ is a positive (negative) definite matrix
$\mathbf{X} \succeq 0 \ (\preceq 0)$	$\mathbf{X}$ is a positive (negative) semi-definite matrix
$\mathbf{A} \circ \mathbf{B}$	Hadamard product, i.e., if $\mathbf{A}, \mathbf{B} \in \mathbb{R}^{n \times m}$, and $\mathbf{C} = \mathbf{A} \circ \mathbf{B} \in \mathbb{R}^{n \times m}$ then $c_{ij} = a_{ij} b_{ij}$, for all $i = \{1, \dots, n\}$ and $j = \{1, \dots, m\}$
$\nabla V(\cdot)$	Gradient of the function V
$D\mathbf{f}(\cdot)$	Jacobian matrix of $\mathbf{f}$, i.e., $[D\mathbf{f}]_{ij} = \frac{\partial f_i}{\partial p_j}$
$(\cdot)^\star$	The super index $\star$ denotes optimality, e.g., $\mathbf{p}^\star$ denotes a Nash equilibrium and $\mathcal{P}^\star$ denotes the optimal partition
$\dot{x}$	Derivative of $x(t)$ with respect to the continuous time, i.e., $\dot{x} = \frac{d}{dt} x(t)$. Moreover, arguments in continuous time are expressed in parenthesis, e.g., $x(t)$, $\mathbf{A}(t)$, and the argument corresponding to the continuous time is mostly omitted throughout this thesis in order to simplify notation
x_k	The sub-index k indicates the discrete time
$x_{k+j\|k}$	Prediction of x made at time instant k for the time instant $k+j$, where $k, j \in \mathbb{Z}_{\geq 0}$. In the argument $k+j\|k$, the first element $k+j$ indicates discrete-time prediction, whereas the second element k indicates the actual discrete time

Nomenclature

$\mathbf{A}$	Adjacency matrix of the graph $\mathcal{G}$
$\tilde{\mathbf{A}}$	Adjacency matrix of the graph $\tilde{\mathcal{G}}$
$\mathbf{A}^{(\mathbf{p})}$	Adjacency matrix of the graph $\mathcal{G}^{(\mathbf{p})}$
$\bar{\mathbf{A}}_i$	Matrix for dynamic model of the ith UAV
$\mathbf{A}_d$	State matrix for a system
$\mathbf{A}_{d,i}$	State matrix for an ith sub-system
a_{ij}	Element in the ith row and jth column in the adjacency matrix $\mathbf{A}$
$\tilde{a}_{ij}$	Element in the ith row and jth column in the adjacency matrix $\tilde{\mathbf{A}}$
$a_{ij}^{(\mathbf{p})}$	Element in the ith row and jth column in the adjacency matrix $\mathbf{A}^{(\mathbf{p})}$
$\mathcal{A}$	Set of UAVs
$\tilde{\mathcal{A}}$	Set of anchor elements
$\bar{\mathbf{B}}_i$	Matrix for dynamic model of the ith UAV
$\mathbf{B}_d$	Control-input matrix for a system
$\mathbf{B}_{d,i}$	Control-input matrix for an ith sub-system
$\mathbf{B}_l$	Disturbance input matrix for a system
$\mathcal{B}$	Set of sink elements in a distribution flow-based network
$\mathbf{C}$	Matrix of relevance factors
c_{ij}	Element in the ith row and jth column in the matrix $\mathbf{C}$
$\mathbf{c}_i$	Vector of the measured position parameters of the ith UAV
C	Energy and water costs
$\tilde{C}_\ell$	Individual cost of the ℓth player
C_S	Energy and water costs with static tuning strategy
C_D	Energy and water costs with dynamic tuning strategy
c_i^d	Measurement of the position parameter d of the ith UAV
$\mathcal{C}_k$	Set of components of the graph $\tilde{\mathcal{G}}_k$ at time instant k
$\mathbf{D}$	Matrix of distances
d	Index to make reference to different populations
$\tilde{d}$	Function of distance between two nodes
$\tilde{d}_{ij}$	Element in the ith row and jth column in the matrix $\mathbf{D}$

$\mathbf{d}$	Vector of the n_d disturbances
$\hat{\mathbf{d}}$	Sequences of disturbances
$\mathcal{D}$	Set of populations
e_{p_ℓ}	Error $p_\ell - p_\ell^\star$
e_{z_ℓ}	error $z_\ell - z_\ell^\star$
E	Lyapunov function
E_V	Lyapunov function
E_1	Storage function (also Lyapunov function)
E_2	Storage function (also Lyapunov function)
$\mathcal{E}$	Set of links/edges of the graph $\mathcal{G}$
$\tilde{\mathcal{E}}$	Set of links/edges of the graph $\tilde{\mathcal{G}}$
$\hat{\mathcal{E}}$	Set of links/edges of the graph $\hat{\mathcal{G}}$
$\mathcal{E}_i$	Set of links/edges of the graph corresponding to the ith topology
$\mathcal{E}_i^j$	Set of links/edges of the graph corresponding to the jth partition in the ith topology
$\mathcal{E}_k^\ell$	Set of links/edges of the graph $\mathcal{G}_k^\ell$
$\mathbf{f}$	Population game, and vector of fitness functions
$\mathbf{f}^d$	dth population game, and vector of fitness functions for the dth population
$\mathbf{f}_i^j$	Population game, and vector of fitness functions for the jth partition in the ith topology
f_i	Fitness function corresponding to the ith strategy
f_i^d	Fitness function corresponding to the ith strategy in the dth population
$\mathcal{F}$	Set of followers
g	Function that determines the index of a partition for a specific node (partitioning context)
$\tilde{g}$	Amount of desired players per coalition (power-index context)
$\bar{g}_i^x$	Upper-bound parameter determining *non-safe* sectors for the ith system state
$\bar{g}_i^u$	Upper-bound parameter determining *non-safe* sectors for the ith control input
$\underline{g}_i^x$	Lower-bound parameter determining *non-safe* sectors for the ith system state
$\underline{g}_i^u$	Lower-bound parameter determining *non-safe* sectors for the ith control input
$\mathcal{G}$	Undirected graph defining possible strategic interactions in a population game
$\tilde{\mathcal{G}}$	Undirected graph corresponding to the system structure (interaction among sub-systems)
$\bar{\mathcal{G}}$	Directed graph representing a distribution flow-based network
$\mathcal{G}^{(\mathbf{p})}$	Graph with adjacency matrix $\mathbf{A}^\mathbf{p}$ in function of $\mathbf{p}$
$\mathcal{G}_i$	Graph corresponding to the ith topology
$\mathcal{G}_i^j$	Graph corresponding to the jth partition in the ith topology
$\mathcal{G}_k^\ell$	The ℓth partition of the graph $\mathcal{G}$ at time instant k

H_p	Prediction horizon in an MPC controller
h_i	Height of the ith tank
$\hat{h}_i$	External benefit for the ith node
$h_{\mathrm{max},\ell}$	Maximum level of the ℓth tank
$\hat{h}_i^\ell$	External benefit for the ith node to change to the ℓth partition
$\breve{h}_i$	Internal benefit for the ith node
J	Cost function in an MPC controller
J_i	The ith cost function in a multi-objective MPC controller
$\tilde{J}_i$	The ith normalized cost function in a multi-objective MPC controller
J_i^ℓ	Cost function of the ith sub-system in an MPC controller
J_i^f	Terminal cost of the ith sub-system in an MPC controller
J_i^{utopia}	The ith value from the vector $\mathbf{J}^{\mathrm{utopia}}$
J_i^{nadir}	The ith value from the vector $\mathbf{J}^{\mathrm{nadir}}$
$\mathbf{J}^{\mathrm{utopia}}$	Utopia point
$\mathbf{J}^{\mathrm{nadir}}$	Nadir point
k	Discrete-time step
K_ℓ	Discharge coefficient of the ℓth tank
$\mathcal{K}$	Set of indices for partitions
$\mathbf{L}$	Laplacian of the graph $\mathcal{G}$
$\mathbf{L}^{(\mathbf{p})}$	Laplacian of the graph $\mathcal{G}^{(\mathbf{p})}$
$\tilde{\mathbf{L}}^{(\mathbf{p})}$	Modified Laplacian in terms of $\mathbf{p}$
l_{ij}	Element in the ith row and jth column in the Laplacian $\mathbf{L}$
$l_{ij}^{(\mathbf{p})}$	Element in the ith row and jth column in the Laplacian $\mathbf{L}^{(\mathbf{p})}$
$\tilde{l}_{ij}^{(\mathbf{p})}$	Element in the ith row and jth column in the Laplacian $\tilde{\mathbf{L}}^{(\mathbf{p})}$
L	Lagrangian function
L_ℓ	Fair costs for the ℓth player determined by the Shapley value
m	Number of partitions
M	Number of agents in a population
$\tilde{M}$	Vehicle mass
$\mathbf{M}_k$	Matrix for communication *key performance index*
n	Number of strategies in a population games
n_u	Number of control inputs
n_x	Number of system states
n_d	Number of disturbances
$n_{u,i}$	Number of control inputs for the ith sub-system
$n_{x,i}$	Number of system states for the ith sub-system
n_i^j	Number of strategies for the jth partition in the ith topology
$\mathrm{NE}(\mathbf{f})$	Set of Nash equilibria for the population game $\mathbf{f}$
$\mathcal{N}_i$	Set of neighbors of the ith node for the graph $\mathcal{G}$
$\tilde{\mathcal{N}}_i$	Set of neighbors of the ith node for the graph $\mathcal{G}$
$\mathcal{O}$	Coalition

$\mathbf{p}$	Vector of portion of agents, or population state
$\mathbf{p}_i$	Reference parameters for the ith UAV local controller
$\mathbf{p}^d$	Vector of portion of agents in the dth population, or the dth population state
$\mathbf{p}_i^j$	Population state for the jth partition in the ith topology
p_i	Portion of agents selecting the ith strategy
p_i^d	Portion of agents selecting the ith strategy in the dth population
$p_{i,\ell}^j$	Portion of agents selecting the ℓth strategy for the jth partition in the ith topology
P_i	Number of partitions for the ith topology
$\mathcal{P}_i$	Set of partitions for the ith topology
$\mathcal{P}_k$	Partitioning at time instant k
$\mathcal{P}_{\text{LMPC}}$	Optimization problem corresponding to a local MPC controller
$\mathcal{P}_{\text{MPC}}$	Optimization problem corresponding to a centralized MPC controller
$\mathcal{P}_{\text{DSD}}$	Optimization problem corresponding to distributed Smith dynamics
$q_{\text{in},\ell}$	Inflow of the ℓth tank
$q_{\text{out},\ell}$	Outflow of the ℓth tank
$\mathbf{Q}_i$	Prioritization matrix of an MPC controller for the ith sub-system
$\tilde{\mathbf{Q}}$	Prioritization matrix of an MPC controller
Q	Total flow resource in a distribution flow-based network
$\mathcal{Q}_{i,k}$	Set of available partitions for the ith node at time instant k
$\mathbf{R}_i$	Prioritization matrix of an MPC controller for the ith sub-system
$\tilde{\mathbf{R}}$	Prioritization matrix of an MPC controller
$\mathbf{r}_i$	Vector of references for the ith sub-system
r_i^d	Reference for the ith UAV for the dth position parameter
$\bar{R}_i^x$	Upper region for system state constraints
$\bar{R}_i^u$	Upper region for control-input constraints
$\underline{R}_i^x$	Lower region for system state constraints
$\underline{R}_i^u$	Lower region for control-input constraints
$\mathcal{R}$	Set of source elements in a distribution flow-based network
$\mathcal{R}_i$	Set of available source elements for the ith non-source element
$\mathcal{S}$	Set of strategies
$\tilde{\mathcal{S}}$	Set of nodes representing sub-systems
$\bar{\mathcal{S}}$	Set of elements in the distribution flow-based network
$\bar{\mathcal{S}}_{\text{st}}$	Set of storage elements in the distribution flow-based network
$\bar{\mathcal{S}}_{\text{ac}}$	Set of actuator elements in the distribution flow-based network
$\bar{\mathcal{S}}_{\text{jo}}$	Set of joint elements in the distribution flow-based network
$\bar{\mathcal{S}}_{\text{si}}$	Set of sink elements in the distribution flow-based network
$\bar{\mathcal{S}}_{\text{so}}$	Set of source elements in the distribution flow-based network
$\mathcal{S}_k^\ell$	Set of nodes in the ℓth partition at time instant k
$\hat{\mathcal{S}}_k^\ell$	External nodes in the ℓth partition at time instant k
$\check{\mathcal{S}}_k^\ell$	Internal nodes in the ℓth partition at time instant k

t	Continuous time instant
T	Number of topologies
TΔ	Tangent space of the simplex set
$\mathcal{T}$	Set of topologies
$\mathbf{U}$	Vector of control inputs within the prediction horizon
$\mathbf{u}$	Vector of control inputs
$\bar{\mathbf{u}}$	Maximum for the vector of control inputs
$\underline{\mathbf{u}}$	Minimum for the vector of control inputs
$\hat{\mathbf{u}}$	Sequence of control inputs
$\mathbf{u}_i$	Vector of control inputs for the ith sub-system
$\mathcal{U}$	Feasible set of control inputs
$\mathbf{V}_k$	Matrix defining differences among number of elements for different partitions at time instant k
$\mathbf{V}_i$	Prioritization matrix for the terminal cost in an MPC controller
V	Potential function
V_c	Characteristic function
v	System state function
v_i	Decoupled system state function for the ith
$\tilde{v}_i$	Coupled system state function for the ith
$v_{ij,k}$	Element in the ith row and jth column in the matrix $\mathbf{V}_k$ at time instant k
$\mathcal{V}$	Set of players
$\mathcal{V}_i^j$	Set of players for the jth partition in the ith topology
$\mathbf{W}_k$	Cost matrix for partitioning
$w_{ij,k}$	Element in the ith row and jth column in the matrix $\mathbf{W}_k$ at time instant k
$\mathcal{W}_k^z$	Set of optimal nodes in the partitioning algorithm at time instant k
$\mathbf{X}$	Vector of system states within the prediction horizon
$\mathbf{X}_r$	Vector of references within the prediction horizon
$\mathbf{x}$	Vector of system states
$\bar{\mathbf{x}}$	Maximum for the vector of system states
$\underline{\mathbf{x}}$	Minimum for the vector of system states
$\hat{\mathbf{x}}$	Sequence of system states
$\mathbf{x}_r$	Vector of references
$\mathbf{x}_i$	Vector of system states for the ith sub-system
$\mathbf{x}_s$	Vector of the desired lower value for system states
$\mathcal{X}$	Feasible set for the system states
z_ℓ	ℓth state in a distribution flow-based network
$z_{\max,\ell}$	Maximum value for the ℓth state in a distribution flow-based network
$\mathbf{z}$	Vector of system states in a distribution flow-based network
$\mathbf{z}_i^j$	Vector of system states in a distribution flow-based network for the jth partition in the ith topology
$\mathbf{z}_{\max,i}^j$	Maximum value for the vector of system states in a distribution flow-based network for the jth partition in the ith topology
$\Delta\mathbf{U}$	Vector of control-input slew rates within the prediction horizon
$\Delta\mathbf{u}$	Vector of control-input slew rates

Δ	Simplex set in a population games
Δ^d	Simplex set for the dth population
Δ°	Simplex set for the density-dependent population game, positive orthant
$\mathrm{int}\Delta$	Interior of the simplex set
$\mathrm{int}\Delta^d$	Interior of the simplex set for the dth population
Δ'	Simplex set without considering positiveness of variables
δ	Reproduction rate function for all strategies
δ_i	Reproduction rate function for the ith strategy
$\varepsilon^\star$	Desired resource-feeding co-relation index
ε^ℓ	Resource-feeding co-relation index
γ	Scalar weight for an objective in the MPC controller
γ_i	Scalar weight for the ith objective in the MPC controller
ς	Auxiliary variable for lower limit system state constraint
$\hat{\varsigma}$	Sequence of auxiliary variable
$\boldsymbol{\mu}$	Lagrange multipliers
$\mu_{r\ell}$	Information provided from sub-system r to sub-system ℓ
ϱ	Revision protocol
$\tilde{\varrho}$	Mean value
ϱ_{ij}	Switch rate from strategy i to strategy j
$\boldsymbol{\lambda}$	Lagrange multipliers
π	Population mass
π^d	dth population mass
π_i^j	Population mass for the jth partition in the ith topology
$\Phi(\ell)$	Shapley value of the player ℓ
$\tilde{\boldsymbol{\Gamma}}$	Diagonal matrix of the active constraints
$\boldsymbol{\Gamma}^u$	Vector of the control-input constraints
$\boldsymbol{\Gamma}^x$	Vector of the system states constraints
$\boldsymbol{\Gamma}_i^j$	Sub-system of the distribution flow-based network for the jth partition in the ith topology
φ	Prioritization vector for the partitioning algorithm
φ_j	Prioritization for the jth objective for the partitioning algorithm
$\boldsymbol{\Sigma}_i^j$	Sub-system of the population dynamics for the jth partition in the ith topology
τ	Sampling time
$\tau^\star$	Maximum resource-feeding index of the network
τ_c	Critical sampling time
τ_i	Network resource-feeding index of the ith node
τ^ℓ	Maximum resource-feeding index per partition
σ_j	jth objective in the partitioning algorithm
η_i	Incentives of the ith to change partition
$\varkappa$	Parameter for stop condition in the partitioning algorithm
$\boldsymbol{\alpha}$	Vector of energy and water costs throughout the prediction horizon
$\boldsymbol{\alpha}_1$	Vector of constant energy costs

$\boldsymbol{\alpha}_{2,k}$ Vector of time-varying water costs
ξ Decision variables in a quadratic programming problem
$\hat{\xi}$ Sequence of the decision variables in a quadratic programming problem within the prediction horizon

Part I
Preliminaries

Chapter 1
Introduction

1.1 Motivation

Large-scale systems have gotten special importance into the engineering field. Last years have witnessed the rise of large-scale systems in parallel with the growth of industry and cities. For instance, the continuous evolution of industrial processes and the necessity of enhancing them have caused the appearance of quite complex systems involving a large number of variables. Similarly, the growth of cities has caused the rise of large-scale systems in charge of supplying the population with resources it needs and that can normally be modeled as a network, e.g., traffic, energy, and water systems. Together with the large-scale systems, the necessity to determine appropriate ways to control and make them perform in a desired manner has also appeared. Finally, both physical and operational constraints should be taken into account, which motivate the use of model predictive control (MPC) since it has been demonstrated to be one of the most used optimization-based control technique that can manage all the required considerations [1, 2]. Therefore, the design of controllers for large-scale systems deals with new challenges. First, the fact that large-scale systems are generally located throughout large geographical areas makes the recollection of measurements and their transmission difficult. In this regard, the communication network required for a centralized control approach might have high associated economical costs, and a large number of links in the communication network is associated to low reliability. Furthermore, the computation of a large amount of data directly implies a high computational burden to manage, process and use them in order to make decisions over the system functioning.

A plausible solution to mitigate the aforementioned issues associated to the control of large-scale systems consists in dividing this type of systems into smaller sub-systems that can be controlled by independent local controllers. From this perspective, local controllers can be associated to local decision makers that use only a portion of information about the whole system, i.e., they only capture information from a sub-system. Nevertheless, the task of identifying the appropriate set of

© Springer International Publishing AG, part of Springer Nature 2019
J. Barreiro-Gomez, *The Role of Population Games in the Design of Optimization-Based Controllers*, Springer Theses,
https://doi.org/10.1007/978-3-319-92204-1_1

sub-systems avoiding the existence of strong dynamical coupling is not a trivial problem and, in many cases, the control requirements and/or the physical and operational constraints do not allow the statement of independent problems that could be solved by smaller independent controllers. As an alternative, distributed approaches have been used, which consist in designing local controllers able to share information among them, i.e., local decision makers that may interact to each other.

The latter analogy incites to look at the distributed control scheme as a scenario where there are several decision makers using limited information and interacting with each other, which motivates the consideration of game theory as a suitable and powerful tool in the study and design of distributed controllers. In addition, one of the most important concepts in the context of game theory is the Nash equilibrium, which describes a situation where no decision maker has incentives to unilaterally change its current decisions. Furthermore, it has been shown that, under certain conditions, there is a close relationship between the Nash equilibrium and the extreme point of a function considering constraints, in other words, there is a relationship between the Nash equilibrium and the solution of a constrained optimization problem. This fact also motivates the purpose of studying the design of distributed optimization-based controllers by using game theoretical approaches.

Instead of designing predictive and game-theory-based controllers as two different control strategies, the motivation in this doctoral thesis is to determine the role of some game-theoretical approaches in the design of optimization-based controllers. In this regard, predictive controllers are designed and studied by exploring two different perspectives. First, these optimization-based controllers are designed partially by using game-theoretical ideas, i.e., considering game theory as a complement to enhance some features of the optimization-based controllers. Secondly, the predictive controllers are completely designed from a game-theoretical perspective.

This doctoral dissertation presents some strategies to complement the design of predictive controllers by means of game-theoretical approaches. As an example, how learning based on population games can be used for the dynamical assignment of prioritization weights for multi-objective cost functions in order to improve the predictive control performance. On the other hand, this thesis addresses the whole design of optimization-based controllers based on game theory. For instance, how to take advantage of the population games features in order to compute control inputs in a distributed manner for engineering problems involving resource allocation. Moreover, the proposed methodologies are implemented in real large-scale case studies to assess the closed-loop performance and their effectiveness by making comparisons with results obtained from centralized control techniques.

1.2 Research Questions

This dissertation is devoted to the design of distributed optimization-based controllers by using game theoretical approaches. The main research goal of this thesis is motivated by the following *key research questions*:

(Q1) Which kind of constrained optimization-based controllers can be designed by using the classical population dynamics and what are the information requirements?

(Q2) How to develop a dynamical tuning methodology for MPC controllers with low computational burden?

(Q3) How to reduce the amount of required information in the evolution of population dynamics?

(Q4) How can the population-games approach be used in the design of distributed optimization-based controllers?

(Q5) How can more coupled constraints be considered with the population-games approach in order to make them suitable for a larger variety of problems in the design of distributed optimization-based controllers?

(Q6) How can the computational burden associated to the computation of the Shapley power index be reduced and how it can be found under a distributed information structure?

(Q7) How can the partitioning of large-scale systems be performed in a distributed manner and how it helps in the design of distributed optimization-based controllers?

(Q8) How can the partitioning of a large-scale system be performed dynamically and how can the population-games approach be used in the design of partitioned optimization-based distributed controllers?

Each one of the aforementioned research questions are addressed throughout this thesis. Question (Q1) consists in identifying the possible control problems that can be solved by using the classical population dynamics considering their information requirements. Therefore, the answer of question (Q1) allows to determine some research opportunities. Moreover, answers to questions (Q2)–(Q8) are the contributions of this doctoral thesis.

1.3 Thesis Outline

This thesis is divided into four parts, i.e.,

(I)   Preliminaries,
(II) The role of games in the design of controllers,
(III) Large-scale systems partitioning in control, and
(IV) Concluding remarks.

The road map of this dissertation is presented in Fig. 1.1, which illustrates the connections among chapters suggesting the reader about their appropriate order. The contents of Chaps. 2–10 are summarized as follows:

Chapter 2: Literature Review and Background

This chapter presents a general literature review covering all the topics discussed in this dissertation. In addition, this chapter introduces the background, preliminary

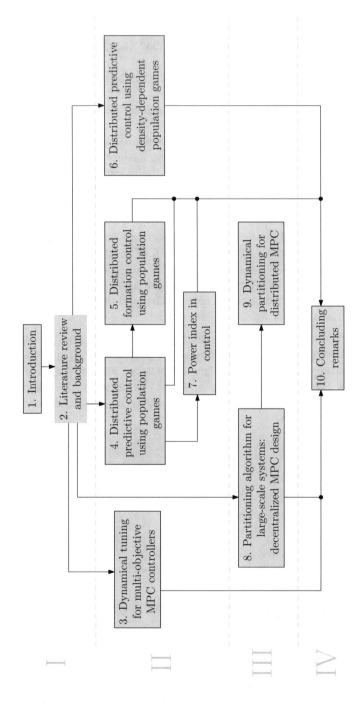

Fig. 1.1 Road map of the thesis. Arrows indicate *read-before* relations

concepts associated to population games and define all the *key research questions.* This chapter answers the *key research question* (Q1) and it is inspired by [3–7].

Chapter 3: Dynamical Tuning for Multi-objective MPC Controllers

This chapter treats the tuning of the weighting parameters of a multi-objective M-PC controller. Normally, the weighting parameters are determined either off-line or by a trial-and-error procedure according to the desired performance of the closed-loop system. However, when the system is affected by disturbances, the appropriate weighting parameters might vary along the time. In this chapter, it is proposed to use a game theoretical approach in order to adjust in an on-line and dynamical manner the prioritization weights depending on the current operational conditions of the system. This chapter answers the *key research question* (Q2) and it is based on [8, 9].

Chapter 4: Distributed Predictive Control Using Population Games

This chapter studies the classical population dynamics presented in this chapter in order to consider strategy-constrained interactions within the population game.[1] As it is presented in this chapter, population dynamics can solve a specific constrained optimization problem (see Problem (2.13)). Additionally, it has been seen that the evolution of the portion of agents depends on the fitness functions of the whole population (see the sum in (2.14), (2.15), (2.16), and (2.17)). This chapter proposes the distributed population dynamics showing a general methodology in order to generate multiple distributed dynamics from different revision protocols, which describe the result and timing about how agents make decisions. Therefore, it is shown how the optimization problem in (2.13) can be solved in a distributed manner even though it contains a coupled constraint. Finally, the novel distributed population dynamics are implemented in the design of a DMPC controller with coupled constraint on the control inputs. This chapter, which answers the *key research question* (Q3) and partially answers the *key research question* (Q4), is based on [10, 11].

Chapter 5: Distributed Formation Control Using Population Games

This chapter shows a formation application of the distributed population dynamics presented in Chap. 4 taking advantage of their properties. It is shown that the stability properties of the equilibrium point of the distributed population dynamics are preserved no matter how the strategy-constrained interactions in the population vary along the time. This fact makes the population-dynamics approach suitable for the formation control when communication limitations are considered. This chapter partially answers the *key research question* (Q4) and it is based on [12].

---

[1]The contributions regarding the distributed population dynamics presented in this chapter have been made and published in [10] with Germán Obando when he was a PhD student at Universidad de los Andes.

Chapter 6: Distributed Predictive Control Using Density-Dependent Population Games

This chapter deals with the issue of considering multiple coupled constraints using population games. To this end, this chapter introduces the density-dependent population dynamics by considering a reproduction rate within the evolution of the portion of agents. Moreover, it is shown how to generate the distributed density-dependent population dynamics and how they can be used in the solution of constrained optimization problems. Furthermore, a DMPC controller is designed based on density games. This chapter answers the *key research question* (Q5) and it is based on [13].

Chapter 7: Power Index in Control

This chapter studies the computation of the Shapley power index and its role in an engineering application regarding the assignment of economical costs for multiple players. One of the main drawbacks associated to the Shapley value is the combinatorial explosion related to its computation. This chapter proposes an alternative manner to compute the Shapley value for a specific characteristic function in order to reduce the computational burden, i.e., from hours to seconds. This reduction in the computational task allows control designers to implement the power index in real-time applications. This chapter answers the *key research question* (Q6) and it is based on [14, 15].

Chapter 8: Partitioning Algorithm for Large-Scale Systems – Sequential DMPC Design

This chapter studies the partitioning of large-scale systems as a powerful tool in the design of distributed optimization-based controllers. This chapter proposes a distributed partitioning algorithm that considers four different objectives and that uses an information-sharing graph. Moreover, it is shown how this partitioning methodology helps in the design of a sequential DMPC controller and it is designed and implemented for a large-scale case study. This chapter answers the *key research question* (Q7) and it is based on [16].

Chapter 9 : Dynamical Partitioning and DMPC

This chapter discusses the dynamical partitioning of a large-scale system along the time. Therefore, it is needed to develop a practical control technique that could cope with such partitioning nature. To this end, it is proposed to use the density-dependent population-games approach introduced in Chap. 6 in combination with the partitioning algorithm introduced in Chap. 8. This chapter answers the *key research question* (Q8).

Chapter 10: Contributions and Concluding Remarks

This chapter draws the concluding remarks of this dissertation. The *key research questions* presented in Sect. 1.2 are also addressed in this chapter.

References

1. Maciejowski J (2002) Predictive control: with constraints. Pearson Education, New York
2. Rawlings JB, Mayne DQ (2009) Model predictive control: theory and design. Nob Hill Publishing, Madison. ISBN 9780975937709
3. Barreiro-Gomez J, Quijano N, Ocampo-Martinez C (2016a) Constrained distributed optimization: a population dynamics approach. Automatica 69:101–116
4. Quijano N, Ocampo-Martinez C, Barreiro-Gomez J, Obando G, Pantoja A, Mojica-Nava E (2017) The role of population games and evolutionary dynamics in distributed control systems. IEEE Control Syst 37(1):70–97
5. Barreiro-Gomez J, Quijano N, Ocampo-Martinez C (2014) Distributed control of drinking water networks using population dynamics: Barcelona case study. In Proceedings of the 53rd IEEE conference on decision and control (CDC). Los Angeles, USA, pp 3216–3221
6. Barreiro-Gomez J, Quijano N, Ocampo-Martinez C (2014) Constrained distributed optimization based on population dynamics. In: Proceedings of the 53rd IEEE conference on decision and control (CDC). Los Angeles, USA, pp 4260–4265
7. Barreiro-Gomez J, Obando G, Riaño-Briceño G, Quijano N, Ocampo-Martinez C (2015) Decentralized control for urban drainage systems via population dynamics: Bogota case study. In: Proceedings of the European control conference (ECC). Linz, Austria, pp 2431–2436
8. Barreiro-Gomez J, Ocampo-Martinez C, Quijano N (2017) Dynamical tuning for MPC using population games: a water supply network application. ISA Trans (to appear) 69(2017):175–186
9. Barreiro-Gomez J, Ocampo-Martinez C, Quijano N (2015) Evolutionary-game-based dynamical tuning for multi-objective model predictive control. In: Olaru S, Grancharova A, Pereira FLobo (eds) Developments in model-based optimization and control. Springer, Berlin, pp 115–138
10. Barreiro-Gomez J, Obando G, Quijano N (2017) Distributed population dynamics: optimization and control applications. IEEE Trans Syst Man Cybern: Syst 47(2):304–314
11. Barreiro-Gomez J, Obando G, Ocampo-Martinez C, Quijano N (2015) Making non-centralized a model predictive control scheme by using distributed smith dynamics. In: Proceedings of the 5th IFAC conference on nonlinear model predictive control. Seville, Spain, pp 501–506
12. Barreiro-Gomez J, Mas I, Ocampo-Martinez C, Peña R. Sánchez, Quijano N (2016) Distributed formation control of multiple unmanned aerial vehicles over time-varying graphs using population games. In: Proceedings of the 55th IEEE conference on decision and control (CDC). Las Vegas, USA, pp 5245–5250
13. Barreiro-Gomez J, Quijano N, Ocampo-Martinez C (2016) Distributed MPC with time-varying communication network: a density-dependent population games approach. In: Proceedings of the 55th IEEE conference on decision and control (CDC). Las Vegas, USA, pp 6068–6073
14. Barreiro-Gomez J, Ocampo-Martinez C, Quijano N, Maestre JM (2017) Non-centralized control for flow-based distribution networks: a game-theoretical insight. J Franklin Inst 354(14):5771–5796
15. Barreiro-Gomez J, Ocampo-Martinez C, Maestre JM, Quijano N (2015) Multi-objective model-free control based on population dynamics and cooperative games. In: Proceedings of the 54th IEEE conference on decision and control (CDC). Osaka, Japan, pp 5296–5301
16. Barreiro-Gomez J, Ocampo-Martinez C, Quijano N (2017) Partitioning for large-scale systems: a sequential dmpc design. In: Proceedings of the 20th IFAC world congress. Toulouse, France, pp 8838–8843

Chapter 2
Literature Review and Background

This chapter presents a literature review related to the main topics treated in this doctoral dissertation. First, a review of model predictive control (MPC) is made focusing on non-centralized schemes, i.e., the architectures for decentralized and distributed MPC controllers. Therefore, some relevant works related to both decentralized and distributed MPC controllers are discussed. Afterwards, a literature review for the tuning issue of the parameters of the MPC controller is introduced. Secondly, the partitioning of large-scale systems is revised, being an essential aspect in the design of non-centralized controllers considering dynamical coupling, information requirements, among others. As a third topic, a review of game-theoretical approaches applied to engineering problems is shown, presenting their versatility in the design of optimization-based controllers. Finally, preliminary concepts regarding population games, which are used throughout the thesis, are presented and some of their relevant features are pointed out.

2.1 Model Predictive Control

Model predictive control (MPC) is one of the most used control strategies in industrial applications because of its versatility to deal with multiple design requirements. The MPC controller is an optimization-based technique that computes an optimal control sequence that minimizes a cost function subject to physical and/or operational constraints at each time instant over a simulation horizon. The latter mentioned feature of the MPC controllers is one of its main advantages with respect to other control strategies.

© Springer International Publishing AG, part of Springer Nature 2019
J. Barreiro-Gomez, *The Role of Population Games in the Design
of Optimization-Based Controllers*, Springer Theses,
https://doi.org/10.1007/978-3-319-92204-1_2

2.1.1 MPC Strategy Description

Even though there exist some mathematical formulations for MPC controllers in continuous time [1], it is more usual to design these kinds of predictive controllers in discrete time [2–5], and by using a state-space model of the system. Therefore, consider a system whose discrete-time model is given by

$$\mathbf{x}_{k+1} = v(\mathbf{x}_k, \mathbf{u}_k, \mathbf{d}_k), \tag{2.1}$$

where $k \in \mathbb{Z}_{\geq 0}$ denotes the discrete time. The vectors $\mathbf{x} \in \mathcal{X} \subseteq \mathbb{R}^{n_x}$, $\mathbf{u} \in \mathcal{U} \subseteq \mathbb{R}^{n_x}$, and $\mathbf{d} \in \mathbb{R}^{n_d}$ correspond to the system states, control inputs and disturbances, respectively. Moreover, the sets $\mathcal{X}$, and $\mathcal{U}$ define the feasible sets according to physical and/or operational constraints for the system states and control inputs. Hence, the function $v : \mathbb{R}^{n_x} \times \mathbb{R}^{n_u} \to \mathbb{R}^{n_x}$ is an arbitrary system state function. Let

$$\hat{\mathbf{u}}_k \triangleq \left(\mathbf{u}_{k|k}, \mathbf{u}_{k+1|k}, \ldots, \mathbf{u}_{k+H_p-1|k} \right)$$

be a feasible control input sequence over a fixed-time prediction horizon denoted by $H_p \in \mathbb{Z}_{>0}$. Moreover, let

$$\hat{\mathbf{x}}_k \triangleq \left(\mathbf{x}_{k+1|k}, \mathbf{x}_{k+2|k}, \ldots, \mathbf{x}_{k+H_p|k} \right)$$

be the system state sequence that is generated when applying the control input sequence $\hat{\mathbf{u}}_k$ to the system (2.1). Moreover, the predictive control approach consists in the solution of an open-loop optimization problem of the following general form:

$$\underset{\mathbf{u}_{k|k},\ldots,\mathbf{u}_{k+H_p-1|k}}{\text{minimize}} \; J(\mathbf{x}_k, \mathbf{u}_k) = J^f \left(\mathbf{x}_{k+H_p|k} \right) + \sum_{j=0}^{H_p-1} J^\ell \left(\mathbf{x}_{k+j|k}, \mathbf{u}_{k+j|k} \right), \tag{2.2a}$$

subject to

$$\mathbf{x}_{k+1+j|k} = v(\mathbf{x}_{k+j|k}, \mathbf{u}_{k+j|k}, \mathbf{d}_{k+j|k}), \; \forall j \in [0, H_p] \cap \mathbb{Z}_{\geq 0}, \tag{2.2b}$$

$$\mathbf{x}_{k+j|k} \in \mathcal{X}, \; \forall j \in [0, H_p] \cap \mathbb{Z}_{\geq 0}, \tag{2.2c}$$

$$\mathbf{u}_{k+j|k} \in \mathcal{U}, \; \forall j \in [0, H_p - 1] \cap \mathbb{Z}_{\geq 0}, \tag{2.2d}$$

where the function $J^\ell : \mathbb{R}^{n_x} \times \mathbb{R}^{n_u} \to \mathbb{R}$ allows to determine the cost throughout the prediction horizon H_p, whereas the function $J^f : \mathbb{R}^{n_x} \to \mathbb{R}$ represents the terminal cost. These functions in (2.2a) should be appropriately selected in order to guarantee the stability of the closed-loop system as it is discussed in [5, 6].

Assuming that the optimization problem in (2.2) is feasible, there is an optimal control input sequence given by

$$\hat{\mathbf{u}}_k^\star \triangleq \left(\mathbf{u}_{k|k}^\star, \mathbf{u}_{k+1|k}^\star, \ldots, \mathbf{u}_{k+H_p-1|k}^\star \right). \tag{2.3}$$

Finally, notice that the receding horizon philosophy only allows to set the first optimal control input from the optimal sequence (2.3) to the system (2.1), i.e., the final control input that is applied to the system is given by

$$\mathbf{u}_{MPC,k}^\star = \mathbf{u}_{k|k}^\star. \tag{2.4}$$

The procedure is repeated at the next iteration, computing a new optimal control sequence and a new final control input. The Algorithm 1 presents the summary of the procedure to compute the MPC law (2.4).

Algorithm 1 General procedure for the computation of the MPC law

---

1: $H_s \leftarrow$ simulation length
2: $H_p \leftarrow$ prediction horizon
3: $k \leftarrow$ initial time
4: $\mathbf{x}_k \leftarrow \mathbf{x}_0 \in \mathbb{R}^{n_x}$ initial condition for the states
5: **for** $k = 1 : H_s$ **do**
6: $\quad$ $\hat{\mathbf{u}}_k^\star \triangleq \left(\mathbf{u}_{k|k}^\star, \mathbf{u}_{k+1|k}^\star, ..., \mathbf{u}_{k+H_p-1|k}^\star \right)$ $\leftarrow$ solve the optimization problem in (2.2)
7: $\quad$ $\mathbf{u}_{MPC,k}^\star = \mathbf{u}_{k|k}^\star$
8: $\quad$ $\mathbf{x}_{k+1} = v(\mathbf{x}_k, \mathbf{u}_{MPC,k}^\star, \mathbf{d}_k)$
9: $\quad$ $k = k + 1$
10: **end for**

2.1.2 Non-centralized MPC Schemes

Even though the MPC controller has been broadly studied by many authors, e.g., [2, 5, 7], there is still a significant interest in studying the design of MPC controllers using non-centralized structures. When coping with the control of a large-scale system, there would be a large number of decision variables and constraints that makes difficult to guarantee the computation of a suitable control input for a given set of performance goals within the established sampling time. Under this scenario, a possible solution is to divide the original problem into smaller and computationally lighter sub-problems, which could be separately solved by using local hardware.

For instance, one of the problems discussed throughout this thesis is related to the appropriate allocation of a limited resource. This kind of problem is appealing since the non-centralized control design should consider a coupled constraint (associated to the resource amount) involving all the control inputs. For a large-scale system composed of several sub-systems, it is common to have a constraint related to the total *energy* resource available for the control inputs, i.e., in real applications, the total energy (resource) demanded by the controllers has an upper bound since the employed resources (e.g., inflows, voltages, forces) are limited. Traditional MPC schemes are capable to overcome this problem by adding that consideration into the set of constraints. Nonetheless, this solution requires the availability of information about the whole system, which implies a centralized control structure that commonly suffers from computational burden issues. Therefore, non-centralized control methods are an alternative. Moreover, different from the design of non-centralized MPC controllers considering a limited resource, i.e., only one coupled constraint in the control inputs, another control problem treated in this dissertation is the design of distributed MPC (DMPC) controllers involving multiple coupled constraints. In this case, the non-centralized design becomes even more challenging since the availability of information should be guaranteed in order to satisfy all the system constraints.

The increasing appearance of complex large-scale systems, e.g., water distribution systems, smart grids, or traffic systems, have motivated the study of non-centralized MPC controllers since some of these systems are not suitable to be controlled with a centralized approach. This fact is motivated by two aforementioned aspects, i.e., communication issues (infrastructure) to collect and transmit data associated to the system states, and computational issues regarding the calculation of control inputs. The problem of obtaining non-centralized control formulations has become a relevant research topic. The process of making controllers non-centralized is normally addressed by dividing the whole system into m different sub-systems and the whole controller into several local smaller controllers as presented in Fig. 2.1a. In [6, 8–10], and more recently in [11], a wide discussion related to non-centralized MPC is developed.

Furthermore, there are several classifications within the non-centralized MPC controllers depending on their architecture and on how different local controllers share information with one another [10]. One of the non-centralized configurations corresponds to decentralized MPC controllers, presented in Fig. 2.1b, where the arrows connecting sub-systems represent the possible dynamical coupling among them. Dashed lines shows that the dynamical coupled might exist or not among sub-systems, e.g., it is possible that two sub-systems are not dynamically coupled. In the decentralized MPC architecture [12], there is a set of local MPC controllers (each one in charge of the control of a sub-system), which do not exchange information to one another. Therefore, in order to implement this control architecture, it is usually assumed that the dynamical coupling among sub-systems is weak. This control configuration has been studied by many authors, e.g., a decentralized MPC algorithm is proposed in [13] for linear large-scale systems and considering constraints over the control inputs. This work proposes to obtain approximated models of sub-systems that represent the global behavior of the system. Another work that considers a linear system structure is presented in [14], where a plug-and-play decentralized MPC controller is proposed. Local controllers only use information from a sub-system and its neighbors, and prior to plug a new sub-system into (or unplug it from) the whole system, it is verified whether or not the modification might affect the closed-loop stability conditions. Stability in this non-centralized architecture has been further analyzed using different methods, e.g., [15, 16]. Regarding applications, a decentralized MPC controller is designed in [17] for an air conditioning system, and in

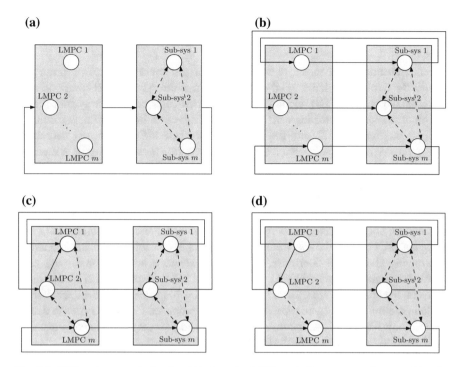

Fig. 2.1 Different non-centralized architectures for MPC controllers: **a** general composition for non-centralized architectures, **b** decentralized architecture, **c** parallel distributed architecture, and **d** sequential distributed architecture. Dashed arrows connecting different sub-systems represent that the dynamical coupling might exist or not, e.g., it is possible that two sub-systems are decoupled

[18] distributed generators in islander mode are controlled by a decentralized MPC controller.

Limitations of the decentralized control configurations have been studied in [19]. Therefore, the performance of the closed-loop system can be enhanced by considering that local MPC controllers can exchange information. Then, these local controllers should be coordinated to obtain a final control input [20]. This modification adding available information among controllers leads to the DMPC architecture as presented in Fig. 2.1c and d. Moreover, DMPC architectures have different sub-categories depending on the manner in which the local MPC controllers share information and also depending on the control objectives they consider. Regarding communication configurations, DMPC controllers are known as parallel when all local controllers communicate simultaneously (see Fig. 2.1c), and they are known as sequential when local controllers exchange information in a sequential manner (see Fig. 2.1d). A comparison among different DMPC schemes is developed in [21]. With respect to the control objectives, DMPC controllers can be cooperative or non-cooperative,[1] i.e., in the non-cooperative DMPC, each controller has its own control objectives, and in the cooperative DMPC, there are common control objectives.

Some authors propose the decomposition of the overall control problem into smaller decoupled problems and the coordination of those individual components in a centralized way [6]. The decomposition of a particular non-linear system for the design of an MPC controller with coupled objectives and constraints is discussed using estimations in [22], and the case in which there are dynamically decoupled sub-systems with coupled objectives and/or constraints is studied in [6]. A methodology of decomposition for a system is also presented in [23], where centralized and distributed approaches are compared in coupled electricity and gas networks. The work presented in [24] considers cooperation approach for DMPC controllers for any finite number of sub-systems. Other distributed strategies propose to share information among different sub-systems at different stages. For instance, in [25], a linear system composed of decoupled sub-systems is controlled by using sequential information sharing. Furthermore, the idea to exchange information among local controllers is also exploited in [26]. Moreover, in [27], the worst case is used in order to ensure the appropriate performance of the closed-loop system, and an accelerated gradient method by using dual decomposition is proposed in [28] in order to design DMPC controllers with faster convergence rates with respect to previously presented duality-based distributed optimization algorithms.

2.1.3 Tuning for MPC Controllers

Consider a multi-objective MPC controller in which the cost function of the optimization problem is composed of n different control objectives, i.e., let the cost function in (2.2a) be as follows:

$$\underset{\mathbf{u}_{k|k},\dots,\mathbf{u}_{k+H_p-1|k}}{\text{minimize}} \quad J(\mathbf{x}_k, \mathbf{u}_k) = J^f\left(\mathbf{x}_{k+H_p|k}\right) + \sum_{i=1}^{n}\sum_{j=0}^{H_p-1} \gamma_i \, J_i^{\ell}\left(\mathbf{x}_{k+j|k}, \mathbf{u}_{k+j|k}\right), \quad (2.5)$$

where the sub-index i is used to differentiate among n cost functions. Moreover, the parameter $\gamma_i \in \mathbb{R}_{\geq 0}$, for all $i = 1, \dots, n$, is a weight that allows to determine a prioritization of the objectives.

Notice that the consideration of multiple control objectives in an MPC controller implies to determine several design parameters, i.e., in (2.5), it is necessary to determine the terminal cost J^f, and to assign values for the prediction horizon H_p, and for the weights γ_i, for all $i = 1, \dots, n$. The task of finding the appropriate values and conditions for the aforementioned parameters is known as the *MPC tuning problem*.

---

[1]The concepts of cooperative and non-cooperative DMPC controllers are omitted due to the fact that it can create confusion with respect to the cooperative and non-cooperative games, which are discussed throughout this doctoral dissertation.

In many cases, the tuning procedure is determined intuitively depending on the engineering application, or the adequate weights are found by a trial-and-error procedure. Furthermore, applications of large-scale nature, the consideration of a large number of constraints and/or the need of including several control objectives make even more complex to determine the appropriate values for the MPC tuning weights. Therefore, the necessity of developing self-tuning methodologies has arisen. Additionally, when having time-varying parameters, disturbances and/or nominal conditions within the system, the appropriate tuning may also vary along the time.

The tuning problem has been discussed by many authors and by using different approaches. A general review about different on-line and off-line tuning approaches for MPC controllers is presented in [29]. An alternative to determine the appropriate tuning of MPC controllers is by matching the MPC performance with the performance of a pre-established controller. For instance, in [30] the tuning of an MPC controller is computed based on a matching to a desired reference controller, then weights are adjusted in order to obtain a behavior close to the performance of the mentioned reference controller. Afterwards, an extension of this approach has been reported in [31]. In [32], the matching to a linear controller is also used to determine the values of the MPC parameters for multiple-input-multiple-output (MIMO) systems. Authors in [33] present a tuning methodology for the weights of an MPC controller in the frequency domain using also control matching. In [34], an automatic tuning strategy is proposed consisting of a controller and a state observer. In [35], a tuning strategy is studied with an optimization algorithm that uses an approximation between both a closed-loop predicted output and the parameters that can be adjusted in the MPC controller. Finally, in [36] an optimal tuning of MPC policies using a stochastic approach is presented.

Other perspectives to solve the problem without the use of a reference model have emerged. For instance, in [37] it is proposed to compute several points of the Pareto front associated to the cost function in a multi-objective MPC controller. Then, a pre-established management point allows to determine the desired value within the Pareto front from which the appropriate tuning weights are determined. In [38], the system output is controlled to maintain it within a region instead of achieving a reference point. Therefore, weights are selected to penalize the output error with respect to a zone for a crude distillation unit. Besides, heuristic directions have also been used to determine the appropriate tuning in an MPC controller as in [39]. Moreover, in [40, 41] the authors use neural networks and fuzzy-based decision making to establish a tuning of an MPC controller, illustrating examples for a mixing tank and for water networks, respectively. Further methods have been explored in the tuning task. In [42], a two-step off-set free tuning procedure is proposed. At first stage, the setup of a nominal MPC loop is made, and then the second step is in charge of adapting the external reference. In [43], a systematic tuning procedure is presented by using multi-objective optimization methods; in [44], a robust tuning problem for a two-degree-of-freedom MPC is presented for single-input-single-output (SISO) system; and authors in [45] have presented a self-tuning of the terminal cost in an economic MPC controller.

2.2 Large-Scale System Partitioning

Large-scale systems are commonly associated to the control of a large number of
states by manipulating also several control inputs. In this regard, controllers must
merge a lot of data and process them in order to compute the appropriate control
inputs and obtain a desired performance for the closed-loop system. Another relevant
aspect that should be considered is that large-scale systems are usually extended
geographically throughout big areas, e.g., flow-based distribution systems, for which
long communication links to transport measurements and control signals are required,
and that might cause additional communication issues and imply economical costs.

The partitioning of large-scale systems appears as a plausible solution in order
to reduce the complexity of the control design, the costs associated to the communication issues and also to reduce the computational complexity. However, the
partitioning task is quite challenging due to the existing dynamical coupling among
elements within the system, imposed coupled constraints, and objective-achievement
warranties, among others.

2.2.1 Partitioning Problem

Consider a large-scale system of the form as in (2.1). Moreover, let the system be
composed of different sub-systems. Therefore, let the set $\tilde{S} = \{1, \ldots, m\}$ represent
the m sub-systems, where the model representing each sub-system $i \in \tilde{S}$ is given by

$$\mathbf{x}_{i,k+1} = v_i(\mathbf{x}_{i,k}, \mathbf{u}_{i,k}, \mathbf{d}_{i,k}) + \sum_{j \in \tilde{S}} \tilde{v}_j(\mathbf{x}_{j,k}, \mathbf{u}_{j,k}, \mathbf{d}_{j,k}), \ \forall i \in \tilde{S}, \qquad (2.6)$$

where $\tilde{v}_j$ represents the existing dynamical coupling between the ith sub-system and
sub-systems $j \in \tilde{S}\setminus\{i\}$. Hence, assume that the system in (2.6) is controlled by an
MPC strategy with an associated optimization problem as presented in (2.2). The
large-scale partitioning consists in determine the appropriate sub-systems $\tilde{S}$ such
that the MPC controller can be performed in a distributed fashion. There are some
general cases that may occur.

• The first case corresponds to a decomposition of the system according to its dynamical coupling. Notice that this partitioning approach can only be applied to specific
system models, e.g., having the condition that the ith sub-system has only inter-
action with other sub-systems in its neighborhood denoted by $\tilde{\mathcal{N}}_i \subset \tilde{S}$. Hence, the
dynamical model of the ith sub-system is given by

$$\mathbf{x}_{i,k+1} = v_i(\mathbf{x}_{i,k}, \mathbf{u}_{i,k}, \mathbf{d}_{i,k}) + \sum_{j \in \{\tilde{\mathcal{N}}_i \cup \{i\}\}} \tilde{v}_j(\mathbf{x}_{j,k}, \mathbf{u}_{j,k}, \mathbf{d}_{j,k}), \ \forall i \in \tilde{S}. \qquad (2.7)$$

Therefore, the decomposition of the system is obtained from its model, i.e., the model imposes conditions over the partitioning. However, the main challenge is to determine a partitioning procedure that could be applied to any large-scale system such that the impact of $\tilde{v}_j$ in (2.6) and (2.7) is minimized or eliminated.

- As a second case, suppose that $\tilde{v}_i = 0$, for all $i \in \tilde{S}$. Therefore, it follows that the dynamical model of the ith sub-system is given by

$$\mathbf{x}_{i,k+1} = v_i(\mathbf{x}_{i,k}, \mathbf{u}_{i,k}, \mathbf{d}_{i,k}), \ \forall i \in \tilde{S}. \tag{2.8}$$

The system in (2.8) shows that all the sub-systems are dynamically decoupled. Nevertheless, notice that it does not mean that the design of a non-centralized controller is not longer challenging. The main reason is that, even though there is not dynamical coupling among sub-systems, it is possible to have coupled constraints in the optimization problem (2.2). A common coupled constraint in engineering problems, related to a resource allocation, is the one involving all the control inputs $\mathbf{u}_{i,k}$, for all $i \in \tilde{S}$.

Finally, it is important to highlight that it is also possible to deal with systems that have a combination of both cases, i.e., systems with dynamical decoupled and coupled sub-systems, and control requirements involving decoupled and coupled constraints in the respective optimization problem.

2.2.2 Some Partitioning Approaches

Most of the partition methods consider a graph representation, i.e., algorithms consider graphs associated to the dynamics of the system (2.1). For instance, a graph-theoretical approach for the decomposition of large-scale systems into a set of interconnected sub-systems is proposed in [46]. However, notice that a partitioning procedure using information about the dynamical model of the system cannot take into account other types of coupling, e.g., constraints involving variables from different sub-systems. Therefore, it is plausible the development of partitioning methodologies that can consider different types of coupling, e.g., dynamical coupling, and coupled constraints among sub-systems.

The partitioning problem has gotten increasing importance in the automatic control community as systems become challenging, and as the requirements and desired closed-loop performance become more strict. Many partitioning proposals focus on specific dynamical systems, or on a particular control strategy. Regarding particular types of systems, the problem of the thermal control for buildings is studied by performing a partition into clusters for decentralized control design in [47]. In [48], a partitioning method is proposed based on capacitor reactive power domains for the control of electric power distribution systems, and in [49], power distribution networks are split into different areas by using a systematic approach in order to control the voltage profile. On the other hand, a method to find an optimal decomposition structure of distributed predictive controllers is presented in [50] by using genetic algorithms. Moreover, the partitioning approach presented in [51] is devoted for the design of decentralized predictive controllers. However, the partitioning methodology is quite general, i.e., the partitioning can be implemented prior to defining the control strategy to apply. Other general partitioning methodologies are presented in [52], where a harmony search algorithm is used, and in [53], where the partitioning is obtained by merging different sub-systems of an initial grouping. Finally, the work presented in [54] discusses a method to determine how information should be shared and the appropriate manner to use it for decentralized, distributed and hierarchical control schemes. There is still a significant interest in the development of general partitioning procedures and methodologies.

As conclusion, the partitioning problem for large-scale systems constitutes a relevant alternative in the design of non-centralized controllers, allowing the identification of multiple sub-systems in an appropriate manner as it has been presented in Sect. 2.1 (see Fig. 2.1). Furthermore, it is important to point out that the partitioning can be addressed in two different ways. The first alternative consists in determining a partitioning based on the dynamical coupling of the whole system, whereas the other possible approach take into account information coupling considering not only the dynamical representation of the system, but also taking into account all the coupled constraints involved in the control design.

2.3 Game Theory

Game theory, which has two different main approaches known as non-cooperative and cooperative, has been applied to many different fields, e.g., economics [55], biology [56, 57], linguistics [58], wireless communication [59], among others. In the last years, game theory has gotten special importance for the design of optimization-based controllers [60], and learning and decision-making algorithms [61]. A general view about the relationship between game theory and distributed control is presented in [62, 63]. It is shown that game theory is quite suitable to achieve global objectives by setting local rules, especially when the engineering problems can be stated as multi-agent systems. Both cooperative and non-cooperative game approaches have been widely used in the design of controllers depending on their control objectives. In some cases, it is more suitable to work with a cooperative perspective when agents can collaborate among them, whereas there are other situations in which it is more appropriate to state the problem as a competition.

2.3.1 Non-cooperative Game Approach

Within the first main branch of game theory, known as the non-cooperative approach, there are other classifications for games depending on how players and strategies are considered, i.e., matricial games, differential games, continuous games, and dynamic games. For instance, in matricial games, players select strategies once and simultaneously from a finite set of possibilities, whereas in continuous games, players can select from an infinite number of strategies [64, 65]. In contrast, under the framework of dynamic games, it is assumed that players that make decisions, are involved in a learning process letting them modify actions based on previous decisions.

Game theory has been implemented for the control of different types of systems. In [66], an overview of the utilization of game theory in signal processing applications is presented. The use of game theory in smart grids can be found in [67, 68]. The work in [69] shows how game theory and decision theory have been applied to multi-agent systems. Finally, a summary of several engineering applications using game theory is presented in [70].

Furthermore, within dynamic games, evolutionary-game theory allows to model the evolution of agents when they interact strategically in a population [71, 72]. This branch has been significantly enriched by the concept of evolutionary-stable strategy (ESS) obtained from assigning stability conditions to a Nash equilibrium [73]. The Nash equilibrium represents a situation in which no player can improve its benefits unilaterally, i.e., there are no players with incentives to change strategy [74]. The result in [73] has been quite used in biology since the concept describes many behaviors found in nature. Afterwards, the evolutionary-stable strategy has been used under a dynamic behavior together with the introduction of the replicator dynamics in [75]. The dynamic-games approach has become a quite useful tool in the design of distributed controllers as it has been presented in [63], and it has been implemented in several fields, e.g., the control of drinking water networks in [76, 77], in the control of wastewater treatment plant in [78], drainage wastewater systems in [79–81], dynamical on-line tuning of predictive controllers in [82], bandwidth allocation in [83], wireless networks in [84], combinatorial optimization in [85], resource allocation problems in [86, 87], cyber security in [88], congestion games in [89], hierarchical frequency control for microgrids in [90], dispatch problem with multiple generators in [91], wind turbines control in [92–94], temperature control in [95], constrained extremum seeking in [96], formation in multi-agent systems in [97], among others.

In the evolution process of the population, each agent makes rational decisions in order to pursue an improvement over its benefits until reaching a scenario where it is not possible to obtain an enhancement by unilaterally making a decision (this situation is given by a Nash equilibrium). Besides, evolutionary-game theory allows to design systems that guarantee convergence to a Nash equilibrium. Additionally, there is a close relationship between the Nash equilibrium with a maximum point in a concave constrained optimization problem due to the fact that, under certain conditions, the Nash equilibrium satisfies the Karush–Kuhn–Tucker (KKT) first-order conditions [71], making evolutionary-game theory a powerful tool to address optimization-based control design. The solution of games can be obtained employing local information. Therefore, if a game framework is applied to address an engineering problem, distributed methodologies emerge. Consequently, the game-theoretical approach becomes a suitable alternative to design distributed controllers.

Population games describe the dynamical process that a population experiences when there is a strategic interaction among the agents that comprise the population [71, 72, 98]. The agents involved in this dynamical process evolve to an equilibrium according to a revision protocol, which establishes the individual decision rules that agents apply to choose the best strategies (i.e., those strategies earning higher payoffs). Population dynamics properties (e.g., passivity [79, 86, 87, 99]) can be exploited to design solutions for a variety of engineering problems. When using population dynamics for solving learning, control, and optimization problems, some elements of the problem are associated with *strategies* that agents in the population can adopt, and other elements are associated with *masses* of agents playing each strategy. This analogy has a direct implication in the information required to implement a solution based on a population dynamics algorithm, since the existing algorithms assume that the population is well-mixed [71, 72] (i.e., any pair of agents playing any pair of strategies can interact with each other). A consequence of the well-mixed population assumption is that the elements of the problem are allowed to interact each other without any constraint (i.e., following a full–information structure). Therefore, classic population dynamics are restricted to be implemented in problems characterized by a centralized information scheme. However, the number of problems that require distributed solutions has increased dramatically in the last few years. In this regard, some approaches have been proposed to model the interaction constraints in a population. For instance, authors in [100, 101] deal with normal-form games and propose a graph-based interaction model, where each node in the graph represents an individual that repetitively plays a symmetric game with its neighbors. However, this approach is not suitable to be applied in the population game framework since, in this type of games, it is preferable to avoid the individuation of players [71]. On the other hand, other approaches aim to apply learning algorithms that are capable to deal with information constraints [102, 103]. Similarly, the authors in [104] modify the well known replicator dynamics model to relax the full-information dependency. They propose a distributed replicator equation in which the evolution of each strategy is only governed by the *neighboring* strategies (according to a given topology).

In addition, one of the main properties of the population games approach is that the evolution of the variables, under population dynamics, evolve inside an invariant set. Under most of the fundamental population dynamics, it is also guaranteed the individual positiveness of each variable. In this regard, population dynamics are able to satisfy a unique coupled constraint. And as it has been mentioned before, dynamics evolve converging to a maximum point of a potential function [71]. The aforementioned features are preserved in a distributed manner by using the distributed version of the population dynamics, which have the same properties as their classical counterpart, i.e., invariance of the simplex set and asymptotic convergence to the equilibrium point [105]. Therefore, distributed population dynamics become a powerful tool to solve constrained optimization problems in a distributed manner, i.e., considering population-interaction constraints (non-complete graphs). Consequently, distributed population dynamics can also be used for the design of distributed optimization-based controllers [63].

The invariance property of the simplex set that represents a coupled constraint and positiveness of variables, and the fact that distributed population dynamics evolve in a non-centralized manner, can be exploited to solve resource allocation problems under non-centralized information-sharing structures. Nevertheless, the necessity to include more constraints, different from the one associated to the simplex set, has gotten special importance to be addressed with population games. For instance in [77], a population dynamics approach, able to solve optimization problems considering multiple constraints, is presented. The authors propose to divide the optimization problems into several smaller sub-problems, and dynamics over each sub-problem feasible region are added in order to achieve an agreement that solves the non-divided optimization problem, allowing also the population size vary along the time.

From a biological perspective, allowing variations of the population size illustrates a special situation in which death and birth, or reproduction rates, are considered as in [106, 107]. Density games have been studied to model a population with reproductive rates. Nevertheless, these dynamics have not been neither deduced from a version of the general dynamics known as *mean dynamics*, and by imposing different rules on revision protocols, nor proposed in a distributed information-sharing fashion. This is because the density-dependent dynamics have been mainly studied in a different context (i.e., biology sciences [106, 107]) from the context where the *mean dynamics* have been mainly applied (e.g., economics [71] and optimization-based control engineering [105]). Besides, the equilibrium points in this type of dynamics have not been related to the solution of constrained optimization problems. Regarding control design, in [87] the dynamics with density dependence presented in [106] have been used for control purposes. Nonetheless, distributed density-dependent population dynamics have not been introduced as a potential tool for centralized/distributed control design.

2.3.2 Cooperative Game Approach

The other main branch of game theory is the cooperative game approach [108, 109]. Cooperative game (or coalitional game) theory studies the conditions and payoff rules for groups of players that form coalitions. Indices of power are alternative manners to solve a game, which are characterized for satisfying a certain system of axioms. Some of these power indices are, among others, the Shapley value, the Banzhaf-Coleman index, or the dictatorial index [110].

A typical solution of a cooperative game is given by the Shapley value [109], which is a power index that assigns a *fair* payoff to each player according to its contribution. Similarly as the non-cooperative games, the cooperative approach has been implemented in different fields, e.g., in politics [110], and economics [111]. Likewise, cooperative games have also been implemented in the engineering context.

Cooperative game theory has been used for example in [112], where a control scheme considering different network topologies is presented. In this and other related works such as [113], the links that compose the network topology are transformed into players of a game and the payoff given by the Shapley value is used as a mean to determine the relevance of the players. Other works that mainly use the Shapley value are [114, 115]. In [114] the Shapley value as a distribution rule is used to guarantee the existence of a Nash equilibrium in any game. In [115] an evolutionary coalitional approach is proposed so that entities can decide in an autonomous manner whether it is profitable or not to make a coalition.

Finally, it is worth to mention that the cooperative-game approach has gotten significant importance in the design of controllers. However, one of the main issues when adopting the cooperative approach consists in the high computational burden to compute a solution [116].

2.3.3 Population Games Concepts

Consider a population composed of a large and finite number of rational agents[2] [71, 117]. These agents select a strategy from the set of n available strategies given by $S = \{1, \ldots, n\}$. The scalar $p_i \in \mathbb{R}_{\geq 0}$ corresponds to the amount of agents, which are selecting the strategy $i \in S$. The population mass is given by $\pi \in \mathbb{R}_{>0}$, then the amount of agents should satisfy that $p_i \leq \pi$, for all $i \in S$. Therefore, the vector $\mathbf{p} \in \mathbb{R}_{\geq 0}^n$ represents the population state or the distribution of agents throughout the strategies, i.e., $\mathbf{p} = [p_1 \quad \cdots \quad p_n]^\top$. Moreover, the set of possible population states is given by a simplex set denoted by

$$\Delta = \left\{ \mathbf{p} \in \mathbb{R}_{\geq 0}^n : \sum_{i \in S} p_i = \pi \right\}, \tag{2.9}$$

and the interior of the simplex set in (2.9) is defined as follows:

$$\mathrm{int}\Delta = \left\{ \mathbf{p} \in \mathbb{R}_{>0}^n : \sum_{i \in S} p_i = \pi \right\}. \tag{2.10}$$

Agents make decisions to select among the different strategies pursuing to increment their benefits. The benefits are determined by a fitness function whose mapping is $f_i : \Delta \to \mathbb{R}$. Depending on the population state, the function f_i takes a population state and returns a real value corresponding to the benefit that the proportion of agents p_i receives for selecting the strategy $i \in S$. The vector of fitness functions for the entire population is denoted by $\mathbf{f}(\mathbf{p})$, whose mapping is given by $\mathbf{f} : \Delta \to \mathbb{R}^n$. Function $\mathbf{f}$ takes a population state and returns a vector of utilities for the population, i.e., $\mathbf{f} = [f_1 \ldots f_n]^\top$.

---

[2]It is assumed that agents are rational in the sense that they are able to make decisions in order to improve their benefits based on current information, i.e., no agent would make a decision that implies a decrement in its current benefits.

Table 2.1 Some revision protocols and their corresponding switch rate

Revision protocol	Switch rate
Pairwise proportional imitation*	$\varrho_{ij}(\mathbf{f}, \mathbf{p}) = p_j[f_j - f_i]_+$
Imitation driven by dissatisfaction**	$\varrho_{ij}(\mathbf{f}, \mathbf{p}) = p_j(K - f_i)$
Pairwise comparison	$\varrho_{ij}(\mathbf{f}, \mathbf{p}) = [f_j - f_i]_+$
Modified pairwise proportional imitation	$\varrho_{ij}(\mathbf{f}, \mathbf{p}) = \frac{[f_j - f_i]_+}{np_i}$
Logit choice	$\varrho_{ij}(\mathbf{f}, \mathbf{p}) = \frac{e^{(\eta^{-1} f_j)}}{\sum_{k \in \mathcal{S}} e^{(\eta^{-1} f_k)}}$

* Notation $[\cdot]_+ = \max(0, \cdot)$ for the *Pairwise proportional imitation protocol*
** K is a large constant in *imitation driven by dissatisfaction* to guarantee that $\varrho_{ij} \geq 0$

In the population, it is assumed that agents are able to migrate from a strategy $i \in \mathcal{S}$ to a strategy $j \in \mathcal{S}$ following a conditional switch rate denoted by $\varrho_{ij}(\mathbf{f}(\mathbf{p}), \mathbf{p})$ that defines the timing and result about how agents make decisions seeking to increment their benefits. The revision protocol function, which determines the conditional switch rates, is presented in Definition 2.1.

Definition 2.1 (*Adapted from* [71]) The revision protocol function is given by a mapping $\varrho : \mathbb{R}^n \times \mathbb{R}^n_{\geq 0} \to \mathbb{R}^{n \times n}_{\geq 0}$ that describes the timing and the results of the agents' decisions in the strategic interaction. Function ϱ takes values corresponding to the population state and its respective benefits, and returns a non-negative matrix representing decisions of agents about switching strategies. ◇

Some of the revision protocols are presented in Table 2.1. According to [71], the evolution process for the portion of agents within a strategic interaction in a large and finite population, considering the way in which agents make decisions, are described by the general dynamics called the *mean dynamics*, i.e.,

$$\dot{p}_i = \sum_{j \in \mathcal{S}} p_j \varrho_{ji}(\mathbf{f}(\mathbf{p}), \mathbf{p}) - p_i \sum_{j \in \mathcal{S}} \varrho_{ij}(\mathbf{f}(\mathbf{p}), \mathbf{p}), \ \forall i \in \mathcal{S}. \qquad (2.11)$$

The dynamics in (2.11) represent the evolution of the proportion of agents p_i, for all $i \in \mathcal{S}$, as a function of a specific revision protocol ϱ. Consequently, the six fundamental population dynamics are generated from (2.11) by setting different revision protocols.

This thesis refers mainly to some specific classes of games such as stable games and full-potential games. These two types of games are presented in Definitions 2.2 and 2.3, respectively.

Definition 2.2 (*Adapted from* [71]) The population game $\mathbf{f} : \Delta \to \mathbb{R}^n$ is a stable game if it satisfies the following condition:

$$(\mathbf{p} - \mathbf{q})^\top (\mathbf{f}(\mathbf{p}) - \mathbf{f}(\mathbf{q})) \leq 0, \ \forall \ \mathbf{p}, \mathbf{q} \in \Delta. \qquad (2.12)$$

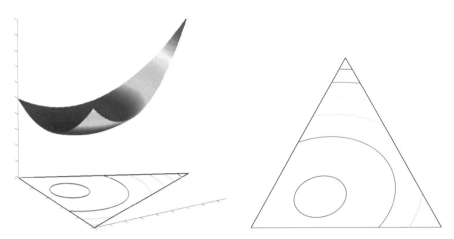

Fig. 2.2 Convex potential function and its projection over the simplex

Alternatively, the condition in (2.12) may be expressed by using the Jacobian matrix of $\mathbf{f}(\mathbf{p})$, i.e., $D\mathbf{f}(\mathbf{p})$. If $\mathbf{f}$ is continuously differentiable, then $\mathbf{f}$ is stable if and only if $\mathbf{z}^\top D\mathbf{f}(\mathbf{p})\mathbf{z} \le 0$, for all $\mathbf{z} \in T\Delta$, and $\mathbf{p} \in \Delta$, where $T\Delta$ is the tangent space of the simplex Δ, defined by $T\Delta = \{\mathbf{z} \in \mathbb{R}^n : \sum_{i \in \mathcal{S}} z_i = 0\}$. Therefore $D\mathbf{f}(\mathbf{p}) \preceq 0$ is a sufficient condition. $\diamond$

There is a close relationship between stable games and a class of population games known as full-potential games defined next.

Definition 2.3 (*Adapted from* [71]) If there exists a continuously differentiable function $V : \mathbb{R}^n_{\ge 0} \to \mathbb{R}$, known as potential function, such that $\mathbf{f}(\mathbf{p}) = \nabla V(\mathbf{p})$, for all $\mathbf{p} \in \mathbb{R}^n_{\ge 0}$, then $\mathbf{f}$ is a full-potential game. More explicitly, if $\frac{\partial V}{\partial p_i}(\mathbf{p}) = f_i(\mathbf{p})$, for all $i \in \mathcal{S}$, and $\mathbf{p} \in \mathbb{R}^n_{\ge 0}$. $\diamond$

Notice that, according to Definitions 2.2 and 2.3, if the potential function $V(\mathbf{p})$ for the full-potential game $\mathbf{f}$ is strictly concave, then $\mathbf{f}$ is a stable game. Figure 2.2 presents a convex potential function $V(\mathbf{p})$ with $\mathbf{p} \in \mathbb{R}^3_{\ge 0}$, whereas Fig. 2.3 presents a concave potential function $V(\mathbf{p})$ that generates a stable game $\mathbf{f}(\mathbf{p})$. Moreover, consider the following constrained optimization problem:

$$\text{maximize}_\mathbf{p} \ V(\mathbf{p}), \tag{2.13a}$$

$$\sum_{i \in \mathcal{S}} p_i = \pi, \tag{2.13b}$$

$$p_i \ge 0. \tag{2.13c}$$

The optimization problem in (2.13) can be solved by using population dynamics by seeking the Nash equilibrium as it has been presented in [71] through the next theorem.

Fig. 2.3 Concave potential function and its projection over the simplex

Theorem 2.1 *If* **f** *is a full-potential game with full-potential function V, then the Nash equilibria satisfy the Karush–Kuhn–Tucker conditions.*

Proof This proof is presented in [71]. □

The Nash equilibrium is a solution in population games since it implies that each rational agent of the population is selecting the best possible strategy against a given population state. Formally, the set of Nash equilibria of a population game is introduced in Definition 2.4.

Definition 2.4 A population state $\mathbf{p}^\star \in \Delta$ is a Nash equilibrium if each used strategy entails the maximum benefit for the proportion of agents that chooses it. Given a population game **f**, let

$$\text{NE}(\mathbf{f}) = \left\{ p_i^\star > 0 \Rightarrow f_i(\mathbf{p}^\star) \geq f_j(\mathbf{p}^\star), \ \forall i, j \in \mathcal{S} \right\}$$

be the set of Nash equilibria. ◇

The six fundamental population dynamics are the replicator, Smith, projection, Brown-Von Neumman-Nash (BNN), Logit choice and best response dynamics. Next, the deduction of some population dynamics are obtained from the mean dynamics, considering the population mass $\pi = 1$, and using some revision protocols [71].

Classical Replicator Dynamics Consider the *pairwise proportional imitation* revision protocol (see Table 2.1) [71]. Therefore, replacing $\varrho_{ij}(\mathbf{f}(\mathbf{p}), \mathbf{p})$ in (2.11) yields

$$\dot{p}_i = p_i \left(f_i(\mathbf{p}) - \sum_{j \in \mathcal{S}} p_j f_j(\mathbf{p}) \right), \quad \forall i \in \mathcal{S}. \tag{2.14}$$

The dynamics presented in (2.14) are the classical replicator dynamics, which have been introduced in [75]. It is worthwhile to point out that the *imitation driven by dissatisfaction* revision protocol also generates the classical replicator dynamics.

Classical Smith Dynamics   Consider the *pairwise comparison* revision protocol (see Table 2.1) [71]. Therefore, replacing $\varrho_{ij}(\mathbf{f}(\mathbf{p}), \mathbf{p})$ in (2.11) immediately yields

$$\dot{p}_i = \sum_{j \in \mathcal{S}} p_j \left[f_i \mathbf{p}) - f_j(\mathbf{p}) \right]_+ - p_i \sum_{j \in \mathcal{S}} \left[f_j(\mathbf{p}) - f_i(\mathbf{p}) \right]_+, \quad \forall\, i \in \mathcal{S}. \quad (2.15)$$

Dynamics in (2.15) are the Smith dynamics, which have been introduced in [118].

Classical Projection Dynamics   Consider the *modified pairwise proportional imitation* revision protocol (see Table 2.1). Therefore, replacing $\varrho_{ij}(\mathbf{f}(\mathbf{p}), \mathbf{p})$ in (2.11) yields

$$\dot{p}_i = f_i(\mathbf{p}) - \frac{1}{n} \sum_{j \in \mathcal{S}} f_j(\mathbf{p}), \quad \forall\, i \in \mathcal{S}. \quad (2.16)$$

The dynamics presented in (2.16) classical projection dynamics [119].

Classical Logit Choice Dynamics   Following the same procedure, the Logit choice dynamics are generated from the mean dynamics and by using the *Logit choice* revision protocol (see Table 2.1) [71], i.e.,

$$\dot{p}_i = \frac{e^{(\eta^{-1} f_i(\mathbf{p}))}}{\sum_{k=1}^{n} e^{(\eta^{-1} f_k(\mathbf{p}))}} - p_i, \quad \forall\, i \in \mathcal{S}, \quad (2.17)$$

where η is a noise level [120].

2.4 Summary

This chapter has presented a general literature review regarding MPC controllers under both decentralized and distributed schemes, and about the tuning issue associated to the design parameters for this type of controllers. Moreover, a review of game theory applied to engineering problems has been presented showing it is a useful tool in the design of optimization-based controllers. Over the end of the review, the system partitioning problem has been discussed, which is an essential aspect for the design of non-centralized controllers. Finally, the preliminary concepts of the classical population games, which are used throughout this thesis, have been presented. For instance, Chap. 3 uses the classical population dynamics for the dynamical tuning of the weighting parameters in the optimization problem behind an MPC controller.

References

1. Wang Liuping (2009) Model predictive control system design and implementation using MATLAB, 1st edn. Springer Publishing Company, Incorporated, Berlin. ISBN 1848823304, 9781848823303
2. Maciejowski J (2002) Predictive control: with constraints. Pearson Education, Berlin
3. Maestre JM Negenborn, RR editors (2014) Distributed model predictive control made easy. Intelligent systems, control and automation: science and engineering, vol 69. Springer, Berlin
4. Ocampo-Martinez C (2010) Model predictive control of wastewater systems. Advances in industrial control, 1st edn. Springer, Berlin. ISBN 978-1-84996-352-7
5. Rawlings JB, Mayne DQ (2009) Model predictive control: theory and design. Nob Hill Publishing, ISBN, p 9780975937709
6. Christofides PD, Scattolini R, Muñoz de la Peña D, Liu J (2013) Distributed model predictive control: A tutorial review and future research directions. Comput Chem Eng 51:21–41
7. Olaru S, Grancharova A, Lobo Pereira F (2015) Developments in model-based optimization and control. Springer, Berlin
8. Camponogara E, Jia D, Krogh B, Talukdar S (2002) Distributed model predictive control. IEEE Control Syst Mag 22(1):44–52
9. Negenborn RR, Maestre JM (2014) Distributed model predictive control: An overview and roadmap of future research opportunities. IEEE Control Syst Mag 34(4):87–97
10. Scattolini R (2009) Architectures for distributed and hierarchical model predictive control - A review. J Process Control 19(5):723–731
11. Mayne D (2014) Model predictive control: Recent developments and future promise. Automatica 50(2014):2967–2986
12. Bemporad A, Barcelli D (2010) Decentralized model predictive control. In: Bemporad A, Heemels M, Johansson M (eds) Networked control systems, vol 406. Lecture notes in control and information sciences. London, Springer, pp 149–178
13. Alessio A, Barcelli D, Bemporad A (2011) Decentralized model predictive control of dynamically coupled linear systems. J Process Control 21:705–714
14. Riverso S, Farina M, Ferrari-Trecate G (2013) Plug-and-play decentralized model predictive control for linear systems. IEEE Trans Autom Control 58(10):2608–2614
15. Magni L, Scattolini R (2006) Stabilizing decentralized model predictive control of nonlinear systems. Automatica 42(2006):1231–1236
16. Raimondo DM, Magni L, Scattolini R (2007) Decentralized model predictive control of nonlinear systems: An input-to-state stability approach. Int J Robust Nonlinear Control 17:1651–1667
17. Elliott MS, Rasmussen BP (2013) Decentralized model predictive control of a multi-evaporator air conditioning system. Control Eng Pract 21(2013):1665–1677
18. Tavakoli A, Negnevitsky M, Muttaqi KM (2016) A decentralized model predictive control for operation of multiple distributed generators in islanded mode. Trans Ind Appl. https://doi.org/10.1109/tia.2016.2616396
19. Cui H, Jacobsen EW (2002) Performance limitations on decentralized control. J Process Control 12:485–494
20. Rawlings JB, Stewart BT (2008) Coordinating multiple optimization-based controllers: New opportunities and challenges. J Process Control 18:839–845
21. Negenborn RR, De Schutter B, Hellendoorn J (2008) Multi-agent model predictive control for transportation networks: serial versus parallel schemes. Appl Artif Intell 21(3):353–366
22. Dunbar W, Murray W (2006) Distributed receding horizon control for multi-vehicle formation stabilization. Automatica 42:549–558
23. Arnold M, Negenborn RR, Andersson G, De Schutter B (2010) Distributed predictive control for energy hub coordination in coupled electricity and gas networks. In: Negenborn RR, Lukszo Z, Hellendoorn H (eds) Intelligent infrastructures. Intelligent systems, control and automation: science and engineering, vol 42. Springer, Netherlands, pp 235–273

24. Ferramosca A, Limon D, Alvarado I, Camacho EF (2013) Cooperative distributed MPC for tracking. Automatica 49(2013):906–914
25. Richards A, How JP (2007) Robust distributed model predictive control. Int J Control 80(9):1517–1531
26. Farina M, Scattolini R (2011) Distributed non-cooperative MPC with neighbour-to-neighbour communication. In: Proceedings of the 18th IFAC world congress, pages 404–409, Milan, Italy,
27. Keviczky T, Borrelli F, Balas G (2004) A study on decentralized receding horizon control for decoupled systems. In: Proceedings of the American control conference (ACC). Boston, USA, pp 4921–4926
28. Giselsson P, Doan MD, Keviczky T, De Schutter B, Rantzer A (2013) Accelerated gradient methods and dual decomposition in distributed model predictive control. Automatica 49:829–833
29. Garriga JL, Soroush M (2010) Model predictive control tuning methods: a review. Ind Eng Chem Res (I&EC) 49:3505–3515
30. Di Cairano S, Bemporad A (2010) Model predictive control tuning by controller matching. IEEE Trans Autom Control 55:185–190
31. Tran QN, Octaviano R, Özkan L, Backx ACPM (2014). Generalized predictive control tuning by controller matching. In: Proceedings of the American control conference (ACC). Portland, USA, pp 4889–4894
32. Shah G, Engell S (2011) Tuning MPC for desired closed-loop performance for MIMO systems. In: Proceedings of the American control conference (ACC). San Francisco, USA, pp 4404–4409
33. Ozkan L, Meijs J, Backx ACPM (2012) A frequency domain approach for MPC tuning. In: Proceedings of the symposium on process systems engineering. Singapore, pp 15–19
34. Waschl H, Alberer D, del Re L (2012) Automatic tuning methods for MPC environments. In Moreno-Díaz R, Pichler F, Quesada-Arencibia A (eds) Computer aided systems theory - EUROCAST 2011: 13th international conference. Revised selected papers, Part II. Springer, Berlin, pp 41–48. ISBN 978-3-642-27579-1
35. Al-Ghazzawi A, Ali E, Nouh A, Zafiriou E (2001) On-line tuning strategy for model predictive controllers. J Process Control 11:265–284
36. Schwartz JD, Rivera DE (2006) Simulation-based optimal tuning of model predictive control policies for supply chain management using simultenuous perturbation stochastic approximation. In: Proceedings of the American control conference (ACC). Minneapolis, Minnesota, USA, pp 14–16
37. Toro R, Ocampo-Martinez C, Logist F, Van Impe J, Puig V (2011) Tuning of predictive controllers for drinking water networked systems. In: Proceedings of the 18th IFAC world congress. Milan, Italy, pp 14507–14512
38. Yamashita AS, Zanin AC, Odloak D (2016) Tuning the model predictive control of a crude distillation unit. ISA Trans 60:178–190
39. Wojsznis W, Gudaz J, Blevins T, Mehta A (2003) Practical approach to tuning MPC. ISA Trans 42:149–162
40. van der Lee JH, Svrcek WY, Young BR (2008) A tuning algorithm for model predictive controllers based on genetic algorithms and fuzzy decision making. ISA Trans 47:53–59
41. Grosso JM, Ocampo-Martinez C, Puig V (2013) Learning-based tuning of supervisory model predictive control for drinking water networks. Eng Appl Artif Intell 26:1741–1750
42. Waschl H, Jogensen JB, Huusom JK, del Re L (2014) A tuning approach for offset-free MPC with conditional reference adaptation. In: Proceedings of the 19th world congress. Cape Town, South Africa, pp 24–29
43. Vallerio M, Impe JV, Logist F (2014) Tuning of NMPC controllers via multi-objective optimisation. Comput Chem Eng 61:38–50
44. He N, Shi D, Wang J, Forbes M, Backstrom J, Chen T (2015) User friendly robust MPC tuning of uncertain paper-making processes. In: Proceedings of the 9th IFAC symposium on advanced control of chemical processes (ADCHEM), vol 48, pp 1021–1026

45. Müller MA, Angeli D, Allgöwer F (2014) On the performance of economic model predictive control with self-tuning terminal cost. J Process Control 24:1179–1186
46. Sezer ME, Šiljak DD, (1986) Nested ε−decompositions and clustering of complex systems. Automatica 22(3):321–331
47. Chandan V, Alleyne A (2013) Optimal partitioning for the decentralized thermal control of buildings. IEEE Trans Control Syst Technol 21(5):1756–1770
48. Kleinberg MR, Miu K, Segal N, Lehmann H, Figura TR (2014) A partitioning method for distributed capacitor control of electric power distribution systems. IEEE Trans Power Syst 29(2):637–644
49. Nayeripour M, Fallahzadeh-Abarghouei H, Waffenschmidt E, Hasanvand S (2016) Coordinated online voltage management of distributed generationusing network partitioning. Electr Power Syst Res 141(2016):202–209
50. Xie L, Cai X, Chen J, Su H (2016) GA based decomposition of large scale distributed model predictive control systems. Control Eng Pract 57(2016):111–125
51. Ocampo-Martinez C, Bovo S, Puig V (2011) Partitioning approach oriented to the decentralised predictive control of large-scale systems. J Process Control 21(2011):775–786
52. Angeline Ezhilarasi G, Swarup KS (2012) Network partitioning using harmony search and equivalencing for distributed computing. J Parallel Distrib Comput 72(2012):936–943
53. Kamelian S, Salahshoor K (2015) A novel graph-based partitioning algorithm for large-scale dynamical systems. Int J Syst Sci 46(2):227–245
54. Núñez A, Ocampo-Martinez C, Maestre JM (2015) De Schutter B (2015) Time-varying scheme for noncentralized model predictive control of large-scale systems. Math Prob Eng 560702:1–17
55. Hidalgo-Gallego S, Núñez-Sánchez R, Coto-Millán P (2016) Game theory and port economics: a survey of recent research. J Econ Surv. https://doi.org/10.1111/joes.12171
56. Hammerstein P, Leimar O (2015) Evolutionary game theory in biology. Handbook of game theory with economic applications 4:575–617
57. Nowak MA, May RM (1992) Evolutionary games and spatial chaos. Nature 359(6398):826–829
58. Jaeger G (2008) Applications of game theory in linguistics. Lang Linguist Compass 2(3):406–421
59. Charilas DE, Panagopoulos AD (2010) A survey on game theory applications in wireless networks. Comput Netw 54(18):3421–3430
60. Giovanini L (2011) Game approach to distributed model predictive control. IET Control Theory Appl 5(15):1729–1739
61. Marden JR, Peyton Young H, Pao LY (2014) Achieving pareto optimality through distributed learning. SIAM J Control Optim 52(5):2753–2770
62. Marden J, Shamma J (2015) Game theory and distributed control. Handbook of game theory with economic applications 4:861–899
63. Quijano N, Ocampo-Martinez C, Barreiro-Gomez J, Obando G, Pantoja A, Mojica-Nava E (2017) The role of population games and evolutionary dynamics in distributed control systems. IEEE Control Syst 37(1):70–97
64. Basar T, Olsder GJ (1999) Dynamic noncooperative game theory, vol 23. SIAM
65. Menache I, Ozdaglar A (2011) Network games: theory, models, and dynamics. Morgan & Claypool Publishers,
66. Bacci G, Lasaulce S, Saad W, Sanguinetti L (2016) Game theory for networks: A tutorial on game-theoretic tools for emerging signal processing applications. IEEE Signal Process Mag 33(1):94–119
67. Saad W, Han Z, Poor HV, Basar T (2012) Game-theoretic methods for the smart grid: An overview of microgrid systems, demand-side management, and smart grid communications. IEEE Signal Process Mag 29(5):86–105, ISSN 1053-5888. https://doi.org/10.1109/MSP.2012.2186410
68. Wang Y, Saad W, Han Z, Poor HV, Baar T (2014) A game-theoretic approach to energy trading in the smart grid. IEEE Trans Smart Grid 5(3):1439–1450. ISSN 1949-3053. https://doi.org/10.1109/TSG.2013.2284664

69. Parsons S, Wooldridge M (2002) Game theory and decision theory in multi-agent systems. Auton Agents Multi-Agent Syst 5(3):243–254
70. Sanchez-Soriano J (2013) An overview on game theory applications to engineering. Int Game Theory Rev 15(03):1340019
71. Sandholm WH (2010) Population games and evolutionary dynamics. MIT Press, Cambridge, Mass
72. Weibull JW (1997) Evolutionary game theory. The MIT Press, London
73. Maynard Smith J, Price G (1973) The logic of animal conflict. Nature 246:15–18
74. Nash JF (1950) Equilibrium points in n-person games. Proc Natl Acad Sci USA 36(1):48–49
75. Taylor PD, Jonker LB (1978) Evolutionary stable strategies and game dynamics. Math Biosci 40(1):145–156
76. Barreiro-Gomez J, Quijano N, Ocampo-Martinez C (2014) Distributed control of drinking water networks using population dynamics: Barcelona case study. In: Proceedings of the 53rd IEEE conference on decision and control (CDC). Los Angeles, USA, pp 3216–3221
77. Barreiro-Gomez J, Quijano N, Ocampo-Martinez C (2016) Constrained distributed optimization: a population dynamics approach. Automatica 69:101–116
78. Barreiro-Gomez J, Quijano N, Ocampo-Martinez C (2015) Distributed resource management by using population dynamics: wastewater treatment application. In: Proceedings of 2nd IEEE Colombian conference on automatic control (CCAC). Manizales, Colombia, pp 1–6
79. Barreiro-Gomez J, Obando G, Riaño-Briceño G, Quijano N, Ocampo-Martinez C (2015) Decentralized control for urban drainage systems via population dynamics: Bogota case study. In: Proceedings of the European control conference (ECC). Linz, Austria, pp 2431–2436
80. Ramirez-Jaime A, Quijano N, Riaño-Briceño G, Barreiro-Gomez J, Ocampo-Martinez C (2016) MatSWMM - an open-source toolbox for designing real-time control of urban drainage systems. Environ Model Softw 83:143–154
81. García L, Barreiro-Gomez J, Escobar E, Téllez D, Quijano N, Ocampo-Martinez C (2015) Modeling and real-time control of urban drainage systems: a review. Adv Water Res 85:120–132
82. Barreiro-Gomez J, Ocampo-Martinez C, Quijano N (2015c) Evolutionary-game-based dynamical tuning for multi-objective model predictive control. In: Olaru S, Grancharova A, Lobo Pereira F (eds) Developments in model-based optimization and control. Springer, Berlin, pp 115–138
83. Poveda J, Quijano N (2012) Dynamic bandwidth allocation in wireless networks using a shahshahani gradient based extremum seeking control. In: Proceedings of the 6th international conference on network games, control and optimization (NetGCooP). Avignon, France, pp 44–50
84. Tembine H, Altman E, El-Azouzi R, Hayel Y (2010) Evolutionary games in wireless networks. IEEE Trans Syst Man Cybern Part B: Cybern 40(3):634–646
85. Bomze I, Pelillo M, Stix V (2000) Approximating the maximum weight clique using replicator dynamics. IEEE Trans Neu Netw 11(6):1228–1241
86. Pashaie A, Pavel L, Damaren CJ (2017) A population game approach for dynamic resource allocation problems. Int J Control 90(9):1957–1972. https://doi.org/10.1080/00207179.2016.1231422
87. Ramirez-Llanos E, Quijano N (2010) A population dynamics approach for the water distribution problem. Int J Control 83:1947–1964
88. Abass AAA, Hajimirsadeghi M, Mandayam NB, Gajic Z (2016) Evolutionary game theoretic analysis of distributed denial of service attacks in a wireless network. In: Proceedings of the 2016 annual conference on information science and systems (CISS). Princeton, USA, pp 36–41. https://doi.org/10.1109/CISS.2016.7460473
89. Sandholm W (2002) Evolutionary implementation and congestion pricing. Rev Econ Stud 69(3):667–689
90. Mojica-Nava E, Macana CA, Quijano N (2014) Dynamic population games for optimal dispatch on hierarchical microgrid control. IEEE Trans Syst Man Cybern: Syst 44(3):306–317

91. Pantoja A, Quijano N (2011) A population dynamics approach for the dispatch of distributed generators. IEEE Trans Ind Electron 58(10):4559–4567
92. Barreiro-Gomez J, Ocampo-Martinez C, Bianchi F, Quijano N (2015d) Model-free control for wind farms using a gradient estimation-based algorithm. In: Proceedings of the European control conference (ECC). Linz, Austria, pp 1516–1521
93. Li N, Marden JR (2013) Designing games for distributed optimization. IEEE J Select Top Signal Process 7(2):230–242. (Special issue on adaptation and learning over complex networks)
94. Marden JR, Ruben SD, Pao LY (2013) A Model-Free Approach to Wind Farm Control Using Game Theoretic Methods. IEEE Trans Control Syst Technol 21(4):1207–1214
95. Obando G, Pantoja A, Quijano N (2014) Building Temperature Control based on Population Dynamics. IEEE Trans Control Syst Technol 22(1):404–412
96. Poveda J, Quijano N (2015) Shahshahani gradient-like extremum seeking. Automatica 58:51–59
97. Barreiro-Gomez J, Mas I, Ocampo-Martinez C, Sánchez R (2016b) Peña, Quijano N (2016) Distributed formation control of multiple unmanned aerial vehicles over time-varying graphs using population games. In: Proceedings of the 55th IEEE conference on decision and control (CDC). Las Vegas, USA, pp 5245–5250
98. Hofbauer J, Sigmund K (1998) Evolutionary games and population dynamics. Cambridge University Press, Cambridge
99. Fox MJ, Shamma JS (2013) Population games, stable games, and passivity. Games 4(4):561–583
100. Berninghaus S, Haller H (2010) Local interaction on random graphs. Games 1(3): 262–285. ISSN 2073-4336. https://doi.org/10.3390/g1030262
101. Alós-Ferrer C, Weidenholzer S (2006) Imitation, local interactions, and efficiency. Econ Lett 93:163–168
102. Boussaton O, Cohen J (2012) On the distributed learning of Nash equilibria with minimal information. In: Proceedings of the 6th international conference on network games, control, and optimization (NetGCooP). Avignon, France, pp 30–37
103. Gharesifard B, Cortes J (2012) Distributed convergence to Nash equilibria by adversarial networks with directed topologies. In: Proceedings of the American control conference (ACC). Montreal, Canada, pp 5881–5886
104. Pantoja A, Quijano N (2012) Distributed optimization using population dynamics with a local replicator equation. In: Proceedings of the 51st IEEE conference on decision and control (CDC). Maui, Hawaii, pp 3790–3795
105. Barreiro-Gomez J, Obando G, Quijano N (2017) Distributed population dynamics: Optimization and control applications. IEEE Trans Syst Man Cybern: Syst 47(2):304–314
106. Cressman R, Křivan V, (2006) Migration dynamics for the ideal free distribution. Am Nat 168(3):384–397
107. Novak S, Chatterjee K, Nowak MA (2013) Density games. J Theor Biol 334(2013):26–34
108. Owen G (1995) Game theory. Academic Press, Cambridge. ISBN 9780125311519
109. Shapley LS (1953) A value for n-person games. Ann Math Stud 28:307–317
110. Owen G, Shapley LS (1989) Optimal location of candidates in ideological space. Int J Game Theory 18(3):339–356
111. Pérez-Castrillo D, Wettstein D (2006) An ordinal shapley value for economic environments. J Econ Theory 127(1):296–308
112. Maestre JM, Muñoz de la Peña D, Jiménez Losada A, Algaba E, Camacho EF (2014) A coalitional control scheme with applications to cooperative game theory. Opt Control Appli Methods 35:592–608
113. Muros Ponce FJ, Maestre JM, Algaba E, Alamo T, Camacho EF (2014) An iterative design method for coalitional control networks with constraints on the Shapley value. In: Proceedings of the 19th IFAC world congress. Cape Town, South Africa, pp 1188–1193
114. Gopalakrishnan R, Marden J, Wierman A (2011) Characterizing distribution rules for cost sharing games. Proceeding of the 5th international conference on network games, control and optimization (NetGCooP). France, Paris, pp 1–4

115. Khan MA, Tembine H, Vasilakos AV (2012) Evolutionary coalitional games: design and challenges in wireless networks. IEEE Wirel Commun 19(2):50–56
116. Deng X, Papadimitriou CH (1994) On the complexity of cooperative solution concepts. Math Oper Res 19(2):257–266
117. Sandholm WH, Dokumaci E, Lahkar R (2008) The projection dynamic and the replicator dynamic. Games Econ Behav 64:666–683
118. Smith MJ (1984) The stability of a dynamic model of traffic assignment-an application of a method of lyapunov. Transp Sci 18(3):245–252
119. Lahkar R, Sandholm WH (2008) The projection dynamic and the geometry of population games. Games Econ Behav 64(2):565–590
120. Ferraioli D (2013) Logit dynamics: a model for bounded rationality. ACM SIGecom Exch 12(1):34–37

Part II
The Role of Games in the Design of Controllers

Chapter 3
Dynamical Tuning for Multi-objective MPC Controllers

This chapter presents a novel methodology for the on-line dynamical tuning of a multi-objective MPC controller based on evolutionary game theory. The contributions presented in this chapter have been published in [2, 3]. The method consists of a normalization of the cost function associated to the optimization problem that the MPC controller solves to determine the optimal control inputs at each time instant, and a population game that fixes the appropriate set of prioritization weights according to a desired region over the Pareto front known as *management region*. Furthermore, the method establishes a convex sum, i.e., the sum of all weights should be equal to one [4]. The population game is solved by using a discrete version of the projection dynamics, which converge to a Nash equilibrium. It is shown that the projection dynamics satisfies the constraint given by the weighted sum, and the stability analysis of the Nash equilibrium under the discrete projection dynamics is formally presented. Some of the previous works mentioned in Chap. 1 related to the tuning problem require either a reference controller or an observer, e.g., [5–7]. Differently, the proposed method in this chapter, based on population games, does not require a reference controller. Moreover, other tuning strategies need to compute several points along the Pareto front in order to select an appropriate prioritization for the local objectives, which implies a high computational burden, e.g., [8]. As an advantage, the proposed method does not require to generate multiple points within the Pareto front associated to the multi-objective cost function in an MPC controller. Furthermore, most of the tuning techniques are static and performed off-line as part of a design procedure. Nevertheless, the proposed tuning methodology can continuously adjust the prioritization of the control objectives to maintain the system operating within the desired *management region*. In order to illustrate the enhancement over the performance of an MPC controller using the dynamical population-games-based tuning, the proposed methodology is applied to a large-scale Water Supply Network (WSN). The results are analyzed and compared with respect to a multi-objective MPC controller with static tuning.

© Springer International Publishing AG, part of Springer Nature 2019
J. Barreiro-Gomez, *The Role of Population Games in the Design
of Optimization-Based Controllers*, Springer Theses,
https://doi.org/10.1007/978-3-319-92204-1_3

3.1 Multi-objective MPC

Consider a system whose dynamics are represented by the following discrete-time state-space model:

$$\mathbf{x}_{k+1} = \mathbf{A}_d \mathbf{x}_k + \mathbf{B}_d \mathbf{u}_k + \mathbf{B}_l \mathbf{d}_k, \tag{3.1}$$

where $k \in \mathbb{Z}_{\geq 0}$ denotes the discrete time step. The vector $\mathbf{x} \in \mathbb{R}^{n_x}$ denotes the system states, $\mathbf{u} \in \mathbb{R}^{n_u}$ denotes the vector of control inputs, and $\mathbf{d} \in \mathbb{R}^{n_d}$ corresponds to the vector of disturbances affecting the system that may be obtained by using a forecasting algorithm as in [9–11], and which is assumed to be known throughout this thesis. The system matrices $\mathbf{A}_d$, $\mathbf{B}_d$, and $\mathbf{B}_l$ are of suitable dimensions. System states and control inputs are constrained because of physical and/or desired operational limits. These constraints are established by defining the following feasible sets:

$$\mathcal{X} \triangleq \left\{ \mathbf{x} \in \mathbb{R}^{n_x} : \mathbf{G}\mathbf{x} \leq \mathbf{g} \right\}, \tag{3.2a}$$

$$\mathcal{U} \triangleq \left\{ \mathbf{u} \in \mathbb{R}^{n_u} : \mathbf{H}\mathbf{u} \leq \mathbf{h} \right\}, \tag{3.2b}$$

where $\mathbf{G}$, $\mathbf{g}$, $\mathbf{H}$, and $\mathbf{h}$ are matrices and vectors of suitable dimensions to represent the constraints for the system states and control inputs, respectively. Let $\hat{\mathbf{u}}_k$ be a sequence of feasible control inputs within a pre-establish prediction horizon denoted by $H_p \in \mathbb{Z}_{>0}$. Similarly, let $\hat{\mathbf{x}}_k$ be the sequence of feasible system states when applying the control input sequence $\hat{\mathbf{u}}_k$ to the system in (3.1). Finally, let $\hat{\mathbf{d}}_k$ be the forecasting of the disturbances. Hence,

$$\hat{\mathbf{u}}_k \triangleq \left(\mathbf{u}_{k|k}, \mathbf{u}_{k+1|k}, \dots, \mathbf{u}_{k+H_p-1|k} \right), \tag{3.3a}$$

$$\hat{\mathbf{x}}_k \triangleq \left(\mathbf{x}_{k+1|k}, \mathbf{x}_{k+2|k}, \dots, \mathbf{x}_{k+H_p|k} \right), \tag{3.3b}$$

$$\hat{\mathbf{d}}_k \triangleq \left(\mathbf{d}_{k|k}, \mathbf{d}_{k+1|k}, \dots, \mathbf{d}_{k+H_p-1|k} \right). \tag{3.3c}$$

Consider that the system (3.1) is controlled by a multi-objective MPC controller with $n \geq 2$ control objectives. The optimization problem behind the MPC controller is stated as follows:

$$\underset{\hat{\mathbf{u}}_k}{\text{minimize}} \; J\left(\mathbf{x}_k, \mathbf{d}_k, \hat{\mathbf{u}}_k \right) = \sum_{i=1}^{n} \gamma_i \, J_i(\mathbf{x}_k, \mathbf{d}_k, \hat{\mathbf{u}}_k), \tag{3.4a}$$

subject to

$$\mathbf{x}_{k+j+1|k} = \mathbf{A}_d \mathbf{x}_{k+j|k} + \mathbf{B}_d \mathbf{u}_{k+j|k} + \mathbf{B}_l \mathbf{d}_{k+j|k}, \quad j \in [0, H_p - 1] \cap \mathbb{Z}_{\geq 0}, \tag{3.4b}$$

$$\mathbf{u}_{k+j|k} \in \mathcal{U}, \quad j \in [0, H_p - 1] \cap \mathbb{Z}_{\geq 0}, \tag{3.4c}$$

$$\mathbf{x}_{k+j|k} \in \mathcal{X}, \quad j \in [1, H_p] \cap \mathbb{Z}_{\geq 0}, \tag{3.4d}$$

where $\mathbf{x}_{k|k} \in \mathbb{R}^{n_x}$ is the current measured system state, and $\gamma_i \in \mathbb{R}_{\geq 0}$, with $i = 1, \ldots, n$, are the n prioritization weights in the cost function $J(\mathbf{x}_k, \mathbf{u})$ satisfying that $\sum_{i=1}^{n} \gamma_i = 1$. Assuming that the optimization problem (3.4) is feasible, its solution is an optimal control input sequence denoted by $\hat{\mathbf{u}}_k^\star$, i.e.,

$$\hat{\mathbf{u}}_k^\star \triangleq \left(\mathbf{u}_{k|k}^\star, \mathbf{u}_{k+1|k}^\star, \ldots, \mathbf{u}_{k+H_p-1|k}^\star \right).$$

Therefore, it follows that the controller applies the first control input from the optimal sequence, which is given by $\mathbf{u}_k^\star \triangleq \mathbf{u}_{k|k}^\star$. Then, after having applied the optimal control input to the system (3.1), a new vector state is measured and the procedure is repeated in order to determine the optimal control sequence from which the optimal control input $\mathbf{u}_{k+1}^\star$ is obtained.

In order to perform the dynamical tuning of the prioritization weights for the previously introduced MPC controller, a discrete-time population-game approach is used. Therefore, one of the classical population dynamics are presented in the following section.

3.2 Classical Projection Dynamics

The projection dynamics are one of the six fundamental population dynamics [12–14], which have been introduced in [15]. These dynamics have been presented in Sect. 2.3.3, which are given by the following differential equation:

$$\dot{p}_i = f_i(\mathbf{p}) - \frac{1}{n} \sum_{j=1}^{n} f_j(\mathbf{p}), \quad \forall i \in \mathcal{S}, \tag{3.5}$$

where $\mathcal{S} = \{1, \ldots, n\}$ is the set of strategies in the population. Then, according to (3.5), the proportion of agents p_i grows as the fitness function $f_i(\mathbf{p})$ is greater than the mean of fitness functions $\frac{1}{n} \sum_{j=1}^{n} f_j(\mathbf{p})$, and decreases otherwise. Alternatively, the projection dynamics in (3.5) can be re-written as follows:

$$\dot{p}_i = \frac{1}{n} \sum_{j=1}^{n} f_i(\mathbf{p}) - \frac{1}{n} \sum_{j=1}^{n} f_j(\mathbf{p}), \quad \forall i \in \mathcal{S},$$

$$\dot{p}_i = \frac{1}{n} \sum_{j=1}^{n} \left(f_i(\mathbf{p}) - f_j(\mathbf{p}) \right), \quad \forall i \in \mathcal{S},$$

$$\dot{\mathbf{p}} = \frac{1}{n} \mathbf{L} \mathbf{f}(\mathbf{p}),$$

where $\mathbf{L}$ corresponds to the Laplacian matrix of a complete graph [16]. The equilibrium point of the projection dynamics (3.5) is achieved when $f_i(\mathbf{p}^\star) =$

$\frac{1}{n}\sum_{j=1}^{n} f_j(\mathbf{p}^\star)$, for all $i \in \mathcal{S}$. This fact implies that at the equilibrium of (3.5), $f_i(\mathbf{p}^\star) = f_j(\mathbf{p}^\star)$, for all $i, j \in \mathcal{S}$, and therefore $\mathbf{p}^\star \in \Delta$ is a Nash equilibrium according to Definition 2.4, i.e., $\mathbf{p}^\star \in \mathrm{NE}(\mathbf{f})$.

For the population-games-based dynamical tuning for multi-objective MPC controllers, it is proposed to use the discrete version of the projection dynamics, which is obtained by using the Euler approximation for a sampling time $\tau \in \mathbb{R}_{>0}$, i.e.,

$$\dot{p}_i \approx \frac{\left(p_{i,k+1} - p_{i,k}\right)}{\tau}.$$

Then,

$$p_{i,k+1} = \tau\left(f_i(\mathbf{p}_k) - \frac{1}{n}\sum_{j=1}^{n} f_j(\mathbf{p}_k)\right) + p_{i,k}, \ \forall i \in \mathcal{S}.$$

Notice that the projection dynamics can be re-written in a compacted manner as follows:

$$\mathbf{p}_{k+1} = \tau\left(\mathbb{I}_n - \frac{1}{n}\mathbb{1}_n\mathbb{1}_n^\top\right)\mathbf{f}(\mathbf{p}_k) + \mathbf{p}_k,$$
$$= \frac{\tau}{n}\mathbf{L}\,\mathbf{f}(\mathbf{p}_k) + \mathbf{p}_k. \tag{3.6}$$

The equilibrium of (3.6) is the same as the equilibrium of (3.5). Then, the equilibrium of (3.6) implies that $f_i(\mathbf{p}^\star) = f_j(\mathbf{p}^\star)$, for all $i, j \in \mathcal{S}$. Prior making the stability analysis of the equilibrium point $\mathbf{p}^\star \in \Delta$, it is shown in Proposition 3.1 that the set of population states Δ is invariant under the discrete projection dynamics (3.6).

Proposition 3.1 *The simplex Δ is an invariant set under the discrete projection dynamics (3.6), i.e., let $\mathbf{p}_0$ be the initial condition of the population state, if $\mathbf{p}_0 \in \Delta$, then $\mathbf{p}_k \in \Delta$, for all $k \in \mathbb{Z}_{\geq 0}$.*

Proof It is desired to prove that $\mathbb{1}_n^\top \mathbf{p}_{k+1} = \mathbb{1}_n^\top \mathbf{p}_k$. Then

$$\mathbb{1}_n^\top \mathbf{p}_{k+1} = \tau\mathbb{1}_n^\top\left(\mathbb{I}_n - \frac{1}{n}\mathbb{1}_n\mathbb{1}_n^\top\right)\mathbf{f}(\mathbf{p}_k) + \mathbb{1}_n^\top\mathbf{p}_k,$$
$$= \tau\mathbb{1}_n^\top\left(\mathbf{f}(\mathbf{p}_k) - \frac{1}{n}\mathbb{1}_n\mathbb{1}_n^\top\mathbf{f}(\mathbf{p}_k)\right) + \mathbb{1}_n^\top\mathbf{p}_k,$$
$$= \tau\left(\mathbb{1}_n^\top\mathbf{f}(\mathbf{p}_k) - \frac{1}{n}\mathbb{1}_n^\top\mathbb{1}_n\mathbb{1}_n^\top\mathbf{f}(\mathbf{p}_k)\right) + \mathbb{1}_n^\top\mathbf{p}_k.$$

Since $\frac{1}{n}\mathbb{1}_n^\top\mathbb{1}_n = 1$, it is obtained that

$$\mathbb{1}_n^\top \mathbf{p}_{k+1} = \tau\left(\mathbb{1}_n^\top\mathbf{f}(\mathbf{p}_k) - \mathbb{1}_n^\top\mathbf{f}(\mathbf{p}_k)\right) + \mathbb{1}_n^\top\mathbf{p}_k.$$

Finally, $\mathbb{1}_n^\top \mathbf{p}_{k+1} = \mathbb{1}_n^\top \mathbf{p}_k$, which completes the proof.   □

The equilibrium point $\mathbf{p}^\star \in \text{NE}(\mathbf{f})$ is asymptotically stable under the discrete projection dynamics (3.6) by selecting appropriately the sampling time τ as stated in Proposition 3.2.

Proposition 3.2 *Let $\mathbf{f}$ be a potential and stable game with potential function $V(\mathbf{p})$, then the equilibrium point $\mathbf{p}^\star \in \Delta$ is asymptotically stable under the discrete projection dynamics (3.6) if the sampling time τ is selected such that the matrix $\Xi(\tau) = \Psi + \frac{\tau}{2}\Psi^\top D\mathbf{f}(\mathbf{p})\Psi$ is positive definite, where $\Psi = \left(\mathbb{I}_n - \frac{1}{n}\mathbb{1}_n\mathbb{1}_n^\top\right) = \frac{1}{n}\mathbf{L}$.*

Proof Since $\mathbf{f}(\mathbf{p}) = \nabla V(\mathbf{p})$, and $\mathbf{f}$ is a stable game, then $V(\mathbf{p})$ is a concave function. Consider the following Lyapunov function candidate:

$$E_k = \frac{V(\mathbf{p}^\star) - V(\mathbf{p}_k)}{\tau},$$

where $E_k > 0$, for all $\mathbf{p} \neq \mathbf{p}^\star$, and $E_k = 0$ for $\mathbf{p} = \mathbf{p}^\star$. It is necessary to show that $\Delta E = E_{k+1} - E_k \leq 0$, i.e.,

$$\Delta E = \frac{V(\mathbf{p}^\star) - V(\mathbf{p}_{k+1}) - V(\mathbf{p}^\star) + V(\mathbf{p}_k)}{\tau},$$

$$= \frac{-V(\mathbf{p}_{k+1}) + V(\mathbf{p}_k)}{\tau}.$$

As in [17], the Taylor expression of $V(\mathbf{p})$ at $\mathbf{p}$ yields

$$V(\mathbf{p}_{k+1}) = V(\mathbf{p}_k) + [\nabla V(\mathbf{p}_k)]^\top \Delta \mathbf{p}_k + \frac{1}{2}[\Delta \mathbf{p}]^\top \nabla^2 V(\mathbf{z}_k)\Delta \mathbf{p}_k,$$

where $\Delta \mathbf{p}_k = \mathbf{p}_{k+1} - \mathbf{p}_k$, and $\mathbf{z}_k$ is a value between $\mathbf{p}_k$, and $\mathbf{p}_{k+1}$. It follows that

$$\Delta E = -\frac{1}{\tau}[\nabla V(\mathbf{p}_k)]^\top \Delta \mathbf{p}_k - \frac{1}{2\tau}[\Delta \mathbf{p}]^\top \nabla^2 V(\mathbf{z}_k)\Delta \mathbf{p}_k. \tag{3.7}$$

Then, replacing from (3.6) the term $\Delta \mathbf{p}$ in (3.7) yields

$$\Delta E = -[\nabla V(\mathbf{p}_k)]^\top \Psi \nabla V(\mathbf{p}) - \frac{\tau}{2}[\nabla V(\mathbf{p})]^\top \Psi^\top \nabla^2 V(\mathbf{z}_k)\Psi \nabla V(\mathbf{p}),$$

$$= -[\nabla V(\mathbf{p}_k)]^\top \left(\Psi + \frac{\tau}{2}\Psi^\top D\mathbf{f}(\mathbf{z}_k)\Psi\right) \nabla V(\mathbf{p}).$$

In conclusion, the equilibrium point $\mathbf{p}^\star \in \Delta$ is asymptotically stable if $\Xi(\tau) = \Psi + \frac{\tau}{2}\Psi^\top D\mathbf{f}(\mathbf{z}_k)\Psi$ is positive definite. In addition, notice that there exists a $\tau \in \mathbb{R}_{>0}$ such that $\Xi \succeq 0$. To verify this fact, ΔE is expressed in terms of the Laplacian $\mathbf{L}$, i.e.,

$$\Delta E = \underbrace{-\frac{1}{n}[\nabla V(\mathbf{p}_k)]^\top \mathbf{L} \nabla V(\mathbf{p})}_{\Delta E^1} \ \underbrace{-\frac{\tau}{2n^2}[\nabla V(\mathbf{p}_k)]^\top \mathbf{L}^\top D\mathbf{f}(\mathbf{z}_k)\mathbf{L}\nabla V(\mathbf{p})}_{\Delta E^2},$$

where the term $\Delta E^1 \le 0$ since it is a quadratic form and $\mathbf{L}$ is positive definite [16], and $\Delta E^2 \ge 0$ since it is a quadratic form and $D\mathbf{f}$ is negative semidefinite according to Definition 2.2. Therefore, there exists a sufficiently small $\tau \in \mathbb{R}_{>0}$ such that $|\Delta E^1| \ge |\Delta E^2|$.   □

Proposition 3.2 requires that the game $\mathbf{f}$ was full potential. Nevertheless, the discrete projection dynamics (3.6) can also be implemented for other types of games. Therefore, Proposition 3.3 presents the stability proof for a game that does not require that the game was full potential, but still stable. Afterwards, it is shown that both results are equivalent for full-potential games.

Proposition 3.3 *Let $\mathbf{f}$ be a stable game, then there exists a sampling time $\tau \in \mathbb{R}_{>0}$ such that the equilibrium point $\mathbf{p}^\star \in \Delta$ is asymptotically stable under the discrete projection dynamics (3.6). The sampling time τ is selected such that $|2(\mathbf{p}_k - \mathbf{p}^\star)^\top \mathbf{f}(\mathbf{p}_k)| > |\tau \mathbf{f}(\mathbf{p}_k)^\top \Psi^\top \Psi \mathbf{f}(\mathbf{p}_k)|$ is satisfied.*

Proof Consider the Lyapunov function $E_k = \frac{1}{\tau}\sum_{i=1}^{n}\left(p_{i,k} - p_i^\star\right)^2$, where $E_k > 0$ for all $\mathbf{p} \ne \mathbf{p}^\star$, and $E_k = 0$ for $\mathbf{p} = \mathbf{p}^\star$. It is necessary to show that $\Delta E = E_{k+1} - E_k \le 0$, i.e.,

$$\Delta E = \frac{1}{\tau}\sum_{i=1}^{n}\{p_{i,k+1}^2 - 2p_{i,k+1}p_i^\star + p_i^{\star2} - p_{i,k}^2 + 2p_{i,k}p_i^\star - p_i^{\star2}\},$$

$$= \frac{1}{\tau}\sum_{i=1}^{n}\{-2p_{i,k+1}p_i^\star + 2p_{i,k}p_i^\star + p_{i,k+1}^2 - p_{i,k}^2\},$$

$$= \frac{1}{\tau}\sum_{i=1}^{n}\{-2p_{i,k+1}p_i^\star + 2p_{i,k}p_i^\star\} + \frac{1}{\tau}\sum_{i=1}^{n}\{p_{i,k+1}^2 - p_{i,k}^2\},$$

$$= -\frac{1}{\tau}\sum_{i=1}^{n}2p_i^\star\left(p_{i,k+1} - p_{i,k}\right) + \frac{1}{\tau}\sum_{i=1}^{n}\left(p_{i,k+1} - p_{i,k}\right)^2 + \frac{1}{\tau}\sum_{i=1}^{n}2p_{i,k}\left(p_{i,k+1} - p_{i,k}\right),$$

$$= \frac{1}{\tau}\sum_{i=1}^{n}2\left(p_{i,k} - p_i^\star\right)\left(p_{i,k+1} - p_{i,k}\right) + \frac{1}{\tau}\sum_{i=1}^{n}\left(p_{i,k+1} - p_{i,k}\right)^2.$$

Replacing the projection dynamics, it follows that

$$\Delta E = \underbrace{2(\mathbf{p}_k - \mathbf{p}^\star)^\top \Psi \mathbf{f}(\mathbf{p}_k)}_{\Delta E^1} + \underbrace{\tau \mathbf{f}(\mathbf{p}_k)^\top \Psi^\top \Psi \mathbf{f}(\mathbf{p}_k)}_{\Delta E^2}.$$

The first term ΔE^1 is re-written as follows:

$$\Delta E^1 = 2(\mathbf{p}_k - \mathbf{p}^\star)^\top \left(\mathbb{I}_n - \frac{1}{n} \mathbb{1}_n \mathbb{1}_n^\top \right) \mathbf{f}(\mathbf{p}_k)$$

$$= 2(\mathbf{p}_k - \mathbf{p}^\star)^\top \mathbf{f}(\mathbf{p}_k) - \frac{2}{n}(\mathbf{p}_k - \mathbf{p}^\star)^\top \mathbb{1}_n \mathbb{1}_n^\top \mathbf{f}(\mathbf{p}_k)$$

$$= 2(\mathbf{p}_k - \mathbf{p}^\star)^\top \mathbf{f}(\mathbf{p}_k) - \frac{2}{n} \underbrace{\left(\mathbf{p}_k^\top \mathbb{1}_n - \mathbf{p}^{\star\top} \mathbb{1}_n \right)}_{0} \mathbb{1}_n^\top \mathbf{f}(\mathbf{p}_k)$$

$$= 2(\mathbf{p}_k - \mathbf{p}^\star)^\top \mathbf{f}(\mathbf{p}_k),$$

then it is concluded that $\Delta E^1 \leq 0$ since $\mathbf{f}$ is stable. On the other hand, $\Delta E^2 = \frac{\tau}{n^2} \mathbf{f}(\mathbf{p}_k)^\top \mathbf{L}^\top \mathbf{L} \mathbf{f}(\mathbf{p}_k)$, and it is concluded that $\Delta E^2 \geq 0$. Finally, there exists a sampling time $\tau \in \mathbb{R}_{>0}$ such that $|\Delta E^1| \geq |\Delta E^2|$.   □

Finding the Sampling Time: A Potential-Game Example

Consider the coordination game given by the following potential function:

$$V(\mathbf{p}) = -\frac{p_1^2}{2} - p_2^2 - \frac{3p_3^2}{2},$$

then, $D\mathbf{f}(\mathbf{p}) = \mathrm{diag}([-1 \quad -2 \quad -3])$. According to Proposition 3.2, the condition for asymptotic stability of the equilibrium point $\mathbf{p}^\star \in \Delta$ is given by

$$\Xi(\tau) = \begin{bmatrix} \frac{2}{3} - \frac{\tau}{2} & \frac{\tau}{6} - \frac{1}{3} & \frac{\tau}{3} - \frac{1}{3} \\ \frac{\tau}{6} - \frac{1}{3} & \frac{2}{3} - \frac{2\tau}{3} & \frac{\tau}{2} - \frac{1}{3} \\ \frac{\tau}{3} - \frac{1}{3} & \frac{\tau}{2} - \frac{1}{3} & \frac{2}{3} - \frac{5\tau}{6} \end{bmatrix}.$$

The conditions over τ to make $\Xi(\tau)$ positive definite are:

$$\frac{2}{3} - \frac{1}{2}\tau > 0, \quad \text{and,} \quad \frac{11}{36}\tau^2 - \frac{2}{3}\tau + \frac{1}{3} > 0.$$

It follows that $\Xi(\tau)$ is positive definite for any $\tau < 0.776$ s, which is the condition to have asymptotic stability of the equilibrium point $\mathbf{p}^\star \in \Delta$ under the discrete projection dynamics (3.6). Figure 3.1 shows the evolution of the proportion of agents $\mathbf{p} \in \Delta$ for the coordination game under the discrete projection dynamics using different sampling times. It can be seen that the system is marginally stable when $\tau = 0.776$ s, validating the condition over τ to have asymptotic stability. Considering Proposition 3.3, it is also possible to find the conditions over the sampling time τ by solving the following problem:

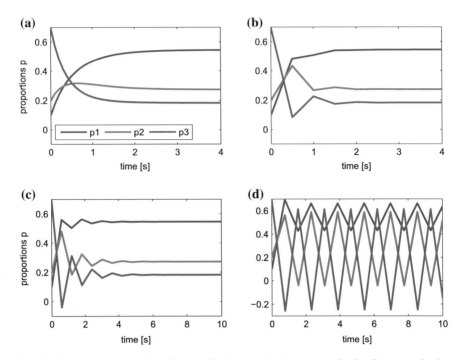

Fig. 3.1 Evolution of proportion of agents for the coordination game under the discrete projection dynamics for four different values of τ. Stable with: **a** $\tau = 0.1 < 0.776$, **b** $\tau = 0.5 < 0.776$, **c** $\tau = 0.6 < 0.776$, and marginally stable with **d** $\tau = 0.776$

$$\underset{\tau \in \mathbb{R}_{>0},\, \mathbf{p} \in \Delta}{\text{minimize}} \quad \tau,$$

$$\text{subject to} \quad 0 \le 2(\mathbf{p} - \mathbf{p}^\star)^\top \mathbf{f}(\mathbf{p}) + \frac{\tau}{n^2}\mathbf{f}(\mathbf{p})^\top \mathbf{L}^\top \mathbf{L}\mathbf{f}(\mathbf{p}),$$

it is the minimum τ such that stability condition is not satisfied with a $\mathbf{p} \in \Delta$. When solving this optimization problem with $\mathbf{f}(\mathbf{p}) = \text{diag}([-1 \quad -2 \quad -3])\mathbf{p}$, it is found that $\tau_c = 0.7762$ is the critical sampling time with $\mathbf{p} = [0.5941 \quad 0.4058 \quad 0]^\top$. This example shows the equivalence between the conditions for τ in Propositions 3.2 and 3.3.

3.3 Proposed Dynamical Tuning Methodology

The proposed dynamical tuning methodology based on population games consists of two different stages. First, it is necessary to normalize the multi-objective cost function, and then the discrete projection dynamics assign permanently the appropriate weights p_i for each one of the control objectives $J_i(\mathbf{x}_k, \mathbf{u})$, for all $i \in \mathcal{S}$. These two main steps of the dynamical tuning methodology are explained next.

3.3.1 Normalization

The cost function (3.4a) has several control objectives, which might depend on different parameters, e.g., one objective depending on the system states in contrast with another objective in function of the control inputs. Furthermore, several objectives (even if they involve the same variables) might have different order of magnitude. Therefore, it is necessary to perform a normalization procedure in order to make a fair comparison among all the control objectives.

Let $\mathbf{x}_i^\star, \mathbf{u}_i^\star$ be the optimal solution of the optimization problem (3.4) considering only the function $J_i(\mathbf{x}_k, \mathbf{u})$, i.e., the solution of (3.4) with weights $\gamma_i = 1$, and $\gamma_j = 0$, for all $j \in \mathcal{S}\backslash\{i\}$. Then, the Utopia point, denoted by $\mathbf{J}^{\text{utopia}} = \left[J_1^{\text{utopia}} \quad \cdots \quad J_n^{\text{utopia}} \right]^\top$, is computed as [18]

$$\mathbf{J}^{\text{utopia}} = \left[J_1(\mathbf{x}_1^\star, \mathbf{u}_1^\star) \quad J_2(\mathbf{x}_2^\star, \mathbf{u}_2^\star) \quad \cdots \quad J_n(\mathbf{x}_n^\star, \mathbf{u}_n^\star) \right]^\top. \tag{3.8}$$

On the other hand, the ith Nadir value is computed as [18]

$$J_i^{\text{nadir}} = \max \left(J_i(\mathbf{x}_1^\star, \mathbf{u}_1^\star), J_i(\mathbf{x}_2^\star, \mathbf{u}_2^\star), \cdots, J_i(\mathbf{x}_n^\star, \mathbf{u}_n^\star) \right), \tag{3.9}$$

where the Nadir point $\mathbf{J}^{\text{nadir}}$ is given by

$$\mathbf{J}^{\text{nadir}} = \left[J_1^{\text{nadir}} \quad J_2^{\text{nadir}} \quad \cdots \quad J_n^{\text{nadir}} \right]^\top. \tag{3.10}$$

Finally, the normalized multi-objective cost function denoted by $\tilde{J}(\mathbf{x}_k, \mathbf{u})$ has the form

$$\tilde{J}(\mathbf{x}_k, \mathbf{u}) = \sum_{i=1}^{n} \tilde{J}_i(\mathbf{x}_k, \mathbf{u}),$$

where each normalized objective is given by

$$\tilde{J}_i(\mathbf{x}_k, \mathbf{u}) = \frac{J_i(\mathbf{x}_k, \mathbf{u}) - J_i^{\text{utopia}}}{J_i^{\text{nadir}} - J_i^{\text{utopia}}}.$$

Having normalized the cost function $J(\mathbf{x}_k, \mathbf{u})$, then the established weights assign a prioritization without being affected by the order of magnitude of each objective. This procedure is illustrated in Fig. 3.2, receiving information from the cost function (that is affected by the forecast of disturbances), prediction model, and constraints.

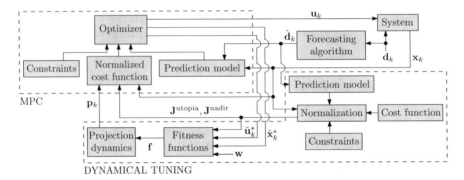

Fig. 3.2 General scheme of the proposed dynamical tuning based on population games

3.3.2 Dynamical Weighting Procedure

Once the cost function has been normalized, it is considered that the prioritization weights at each control objective $J_i(\mathbf{x}_k, \mathbf{u})$ are given by a time-varying parameter $p_{i,k}$, for all $i \in \mathcal{S}$. Hence, the normalized optimization problem behind the MPC controller is formulated as follows:

$$\underset{\hat{\mathbf{u}}_k}{\text{minimize}} \sum_{i=1}^{n} p_{i,k} \tilde{J}_i(\mathbf{x}_0, \hat{\mathbf{u}}_k), \tag{3.11a}$$

subject to:

$$\mathbf{x}_{k+j+1|k} = \mathbf{A}_d \mathbf{x}_{k+j|k} + \mathbf{B}_d \mathbf{u}_{k+j|k} + \mathbf{B}_l \mathbf{d}_{k+j|k}, \quad j \in [0, H_p - 1] \cap \mathbb{Z}_{\geq 0}, \tag{3.11b}$$

$$\mathbf{u}_{k+j|k} \in \mathcal{U}, \quad j \in [0, H_p - 1] \cap \mathbb{Z}_{\geq 0}, \tag{3.11c}$$

$$\mathbf{x}_{k+j|k} \in \mathcal{X}, \quad j \in [1, H_p] \cap \mathbb{Z}_{\geq 0}, \tag{3.11d}$$

where $\mathbf{p}_k = \begin{bmatrix} p_{1,k} & \cdots & p_{n,k} \end{bmatrix}^\top$, satisfying the constraint $\sum_{i=1}^{n} p_{i,k} = 1$. The unitary value at the constraint of the weights sum is associated to the population mass that defines the simplex set Δ in the population game. Notice that weights are able to vary dynamically since the disturbances in the system (3.1) might also vary along the time. To overcome this issue, the discrete projection dynamics (3.6) are implemented. The fitness functions $f_i(p_{i,k})$, for all $i \in \mathcal{S}$, are chosen to be dependent of the current value of each control objective $\tilde{J}_i(\hat{\mathbf{x}}_k^\star, \hat{\mathbf{u}}_k^\star)$ representing each strategy, i.e.,

$$f_i(p_{i,k}) = w_i \tilde{J}_i(\hat{\mathbf{x}}_k^\star, \hat{\mathbf{u}}_k^\star), \tag{3.12}$$

where w_i, for all $i \in \mathcal{S}$, assigns a prioritization that defines a *management region* in the Pareto front as has been presented in [2]. Besides, these terms w_i, for all $i \in \mathcal{S}$, do not appear in the optimization problem of the MPC, and should not be confused with the weights of the cost function (3.11a) in the MPC controller, which are denoted by p_i, for all $i \in S$. Furthermore, notice that the potential function $V(\mathbf{p})$ is unknown for this game. The function $V(\mathbf{p})$ would represent the Pareto front in function of the all possible assigned prioritization (see Remark 3.1).

The differences between the *management region* and the static weights in the multi-objective optimization problem are discussed. To do so, consider a simple and general optimization problem given by

$$\underset{\mathbf{z}}{\text{minimize}} \ J(\mathbf{z}) = p_1 J_1(\mathbf{z}) + p_2 J_2(\mathbf{z}), \qquad (3.13a)$$

subject to

$$\mathbf{Hz} \leq \mathbf{h} + \mathbf{c}, \qquad (3.13b)$$

where $\mathbf{z} \in \mathbb{R}^n$ is the decision variable, and $\mathbf{H} \in \mathbb{R}^{l \times n}$ is a constant matrix with suitable dimension. The values $p_1, p_2 \in \mathbb{R}_{\geq 0}$ establish a static prioritization for the objectives $J_1(\mathbf{z})$ and $J_2(\mathbf{z})$, respectively. The vector $\mathbf{h} \in \mathbb{R}^l$ is a constant component in the constraint, whereas the vector $\mathbf{c} \in \mathbb{R}^l$ is a time-varying component. For instance, the time-varying value of the vector $\mathbf{c} \in \mathbb{R}^l$ may be associated to a disturbance $\mathbf{d} \in \mathbb{R}^l$ involved in a constraint in the optimization problem of an MPC controller.

First, suppose that $\mathbf{c} = \mathbf{c}_1$ in (3.13b), and let $\mathbf{c}_1 \in \mathbb{R}^l$ be a vector of arbitrary entries. For this case, suppose that the obtained Pareto front is the one presented in Fig. 3.3a, and its normalized Pareto front is the one presented in Fig. 3.3b. This figure shows an example in which the management region is given by $w_1 = w_2 = 0.5$, and shows the solution for the optimization problem when static weights in the multi-objective functions are assigned as $p_1 = p_2 = 0.5$ to objectives $J_1(\mathbf{z})$, and $J_2(\mathbf{z})$, respectively. Notice the difference between the selection of the *management region* and the assignment of the weights in the cost function.

Now, suppose that $\mathbf{c}$ in (3.13b) varies, e.g., $\mathbf{c} = \mathbf{c}_2$, where the entries of $\mathbf{c}_1$, and $\mathbf{c}_2$ are near values, i.e., $\mathbf{c}_1 - \mathbf{c}_2 \approx \mathbf{0}$. In this case, the Pareto front varies. Suppose that the new Pareto front is the one obtained in Fig. 3.3c, with its corresponding normalized front presented in Fig. 3.3d. When making this modification over $\mathbf{c}$, the solution of the optimization problem for the weights $p_1 = p_2 = 0.5$ changes dramatically over the Pareto front (this fact illustrates the effect when the disturbances, denoted by $\mathbf{d}$, vary in the optimization problem (3.11)). However, notice that the *management region* is still defined as a region where the objective functions have a equitable value for the particular case $w_1 = w_2 = 0.5$.

When the *management region* is defined, the dynamical tuning strategy is in charge of finding the proper weights $\tilde{p}_1$, and $\tilde{p}_2$ in the normalized cost function, such that the solution lies inside the *management region*. This philosophy is different from the static tuning strategy where the weights are determined previously. The process to assign dynamically the tuning weights is performed by using the population dynamics.

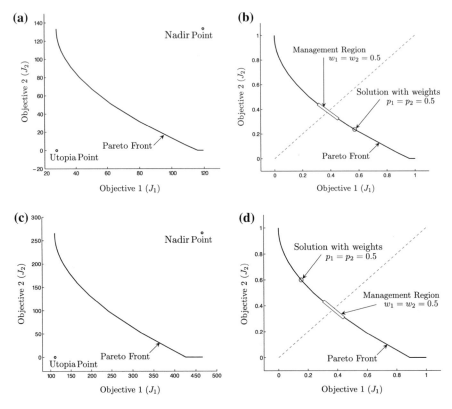

Fig. 3.3 Comparison between the *management region* and the optimization prioritizing weights

Assumption 3.1 The fitness function $f_i(p_i)$ is a strictly decreasing function with respect to p_i. It is expected that the value of the objective $\tilde{J}_i(\hat{\mathbf{x}}_k^\star, \hat{\mathbf{u}}_k^\star)$ decreases as bigger weight $p_{i,k}$ is assigned to it when solving the corresponding optimization problem. $\diamondsuit$

Remark 3.1 Propositions 3.2 and 3.3 have shown that there exists a sampling time $\tau \in \mathbb{R}_{>0}$ such that the equilibrium point $\mathbf{p}^\star \in \Delta$ is asymptotically stable under the discrete projection dynamics. Moreover, in order to find the critical τ_c, it is necessary either to compute the Jacobian $D\mathbf{f}(\mathbf{p})$ or to know the equilibrium point $\mathbf{p}^\star \in \Delta$. For the dynamical tuning application, none of these data are available since there is not a function describing the Pareto front depending on the assigned prioritization in the cost function (3.11a), and the equilibrium point might vary along the time because of the time-varying disturbance affecting the system. However, there exists a sufficiently small τ to guarantee stability according to Proposition 3.3 since the game $\mathbf{f}$ is stable. For the tuning application, a sampling time $\tau = 0.15 < \tau_c$ has been selected. $\diamondsuit$

The dynamical adjustment of the weights is presented in Fig. 3.2. The fitness functions are determined by using information from the normalized cost function and the weights that determine the *management region*. Thus, the discrete projection dynamics compute the appropriate prioritization of the normalized cost function in the MPC controller. A detailed procedure to implement the population-games-based dynamical tuning for multi-objective MPC is presented in Algorithm 2. Notice that it is necessary to compute the normalization at each iteration since the disturbances vary along the time.

Algorithm 2 Dynamical tuning based on population games for multi-objective MPC.

1: $H_s \leftarrow$ simulation length
2: $H_p \leftarrow$ prediction horizon
3: $n \leftarrow$ number of objectives
4: $k \leftarrow$ initial time
5: $\mathbf{x}_k \leftarrow \mathbf{x}_0 \in \mathbb{R}^{n_x}$ initial condition for the states
6: $\mathbf{p}_k \leftarrow \mathbf{p} \in \mathbb{R}^n_{\geq 0}$ initial condition for the proportion
7: **for** $k = 1 : H_s$ **do**
8:     **for** $i = 1 : n$ **do**
9:         $\mathbf{u}_i^\star \leftarrow \arg \min_{\hat{\mathbf{u}}} J_i(\mathbf{x}, \mathbf{u})$ with constraints
10:        $J_i^{\text{utopia}} \leftarrow J_i(\mathbf{x}_i^\star, \mathbf{u}_i^\star)$
11:    **end for**
12:    **for** $j = 1 : n$ **do**
13:        $J_j^{\text{nadir}} \leftarrow \max \left(J_j(\mathbf{x}_1^\star, \mathbf{u}_1^\star), \cdots, J_j(\mathbf{x}_n^\star, \mathbf{u}_n^\star) \right)$
14:    **end for**
15:    $\hat{\mathbf{x}}_k^\star, \hat{\mathbf{u}}_k^\star \leftarrow \arg \min_{\hat{\mathbf{u}}} \sum_{i=1}^n p_{i,k} \tilde{J}_i(\mathbf{x}, \mathbf{u})$ with constraints
16:    $\mathbf{u}_k^\star \leftarrow \mathbf{u}_{k|k}^\star \in \mathbb{R}^{n_u}$ optimal control input
17:    **for** $i = 1 : n$ **do**
18:        $f_i(p_i) \triangleq f_i(p_{i,k}) \leftarrow w_i \tilde{J}_i(\hat{\mathbf{x}}_k^\star, \hat{\mathbf{u}}_k^\star)$
19:    **end for**
20:    $\mathbf{p}_{k+1} = \tau \left(\mathbb{I}_n - \frac{1}{n} \mathbb{1}_n \mathbb{1}_n^\top \right) \mathbf{f}(\mathbf{p}_k) + \mathbf{p}_k$
21:    $\mathbf{x}_{k+1} = \mathbf{A}_d \mathbf{x}_k + \mathbf{B}_d \hat{\mathbf{u}}_k^\star + \mathbf{B}_l \mathbf{d}_k$
22: **end for**

3.3.3 Case Study: Barcelona Water Supply Network

This section introduces the Barcelona water supply network (BWSN) (see [19] for further details regarding this case study) and the design of a multi-objective MPC controller. In order to illustrate the performance of the aforementioned multi-objective MPC controller with a dynamical tuning based on population games, the proposed on-line tuning methodology is implemented in a large-scale water supply network. Furthermore, the performance of the MPC controller with dynamical tuning is compared to the performance obtained by using a conventional static tuning. Figure 3.4 shows a representative portion of the BWSN that is composed of 17 tanks, 26 pumps, 35 valves, nine water sources, 25 water demands, and 11 mass-balance nodes. The dynamical model of the system is given by the following expressions:

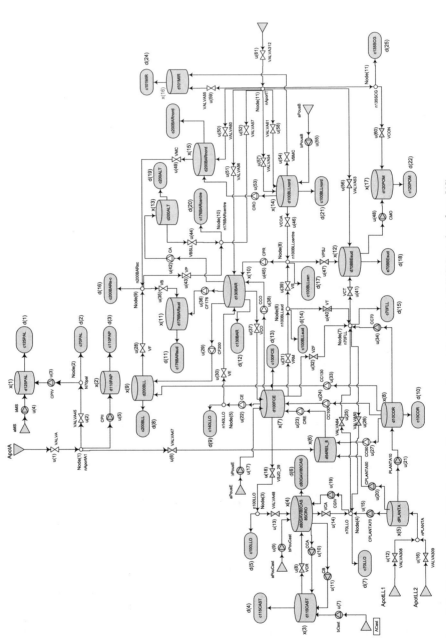

Fig. 3.4 Case study. Topology of the 17 tanks of the Barcelona water supply network (BWSN) (taken from [1])

$$\mathbf{x}_{k+1} = \mathbf{A}_d\mathbf{x}_k + \mathbf{B}_d\mathbf{u}_k + \mathbf{B}_l\mathbf{d}_k, \tag{3.14a}$$

$$0 = \mathbf{E}_u\mathbf{u}_k + \mathbf{E}_d\mathbf{d}_k, \tag{3.14b}$$

where $\mathbf{x} \in \mathbb{R}^{n_x}$ is the vector of $n_x = 17$ system states corresponding to the tank volumes, $\mathbf{u} \in \mathbb{R}^{n_u}$ is the vector of $n_u = 61$ control inputs, and $\mathbf{d} \in \mathbb{R}^{n_d}$ is the vector of $n_d = 25$ time-varying water demands. The water demands are considered to be disturbances to the system, which have a periodicity of 24 hours with a mean value, and a nominal amplitude [10]. The constraints given by the 11 mass-balance nodes are described by (3.14b), i.e., the constraint represents the static water balance for all the nodes in the WSN. Matrices $\mathbf{A}_d$, $\mathbf{B}_d$, $\mathbf{B}_l$, $\mathbf{E}_u$, and $\mathbf{E}_d$ are obtained according to the control-oriented modeling described in [20].

3.3.4 Management Criteria

The MPC controller is designed considering a cost function with multiple objectives. These objectives for the BWSN are established by a *management criteria* considering the following three aspects:

- Economic operation, i.e., $J_1(\mathbf{u}_k) \triangleq \left|\left(\alpha_1 + \alpha_{2,k}\right)^\top \mathbf{u}_k\right|$, where α_1 represents the time-invariant costs associated to the water resource, and $\alpha_{2,k}$ represents the time-varying electricity costs associated to the operation of valves and pumps.
- Smoothness operation, i.e, $J_2(\mathbf{u}_k) \triangleq \|\Delta\mathbf{u}_k\|^2$, where $\Delta\mathbf{u}_k = \mathbf{u}_k - \mathbf{u}_{k-1}$.
- Safety operation, i.e., considering the constraint $\mathbf{x}_k \geq \mathbf{x}_s - \varsigma_k$, for all k, with $\mathbf{x}_s \in \mathbb{R}^{n_x}$ being the vector of safety volumes for all the tanks, and $\varsigma_k \in \mathbb{R}^{n_x}_{\geq 0}$. The third objective is given by $J_3(\varsigma_k) \triangleq \|\varsigma_k\|^2$.

It is important to clarify that the prioritization of objectives, which is determined by the company in charge of the management of the network, is already known. In fact, the prioritization of these aforementioned objectives is commonly used in the design of controllers using a static tuning [9, 19]. In this particular case study, and according to the company in charge of the system, the most important objective is the minimization of the economical costs, i.e., $J_1(\mathbf{u}_k)$. The second most important objective is the one related to the safety volumes, i.e., $J_3(\varsigma_k)$. Finally, the *less* important control objective is related to the smooth operation, i.e., $J_2(\mathbf{u}_k)$. This prioritization order of the objectives in the cost function should be satisfied in case of both static and dynamical tuning.

Figure 3.5 shows the trend of the normalized functions $\tilde{J}_i(\hat{\mathbf{x}}_k^\star, \hat{\mathbf{u}}_k^\star)$, for all $i = 1, 2, 3$. It can be seen that these functions are decreasing with respect to the weight p_i. This is because it is expected to get a smaller value from the minimization problem as more prioritization is assigned (see Assumption 3.1).

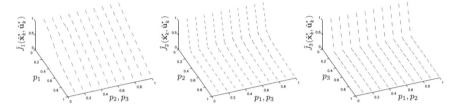

Fig. 3.5 Behavior of the trend of the normalized functions $\tilde{J}_i(\hat{\mathbf{x}}_k^\star, \hat{\mathbf{u}}_k^\star)$, for all $i = 1, 2, 3$. All these functions are decreasing, and in particular, $\tilde{J}_2$ and $\tilde{J}_3$ decrease with a very small slope

3.3.5 Scenarios

In order to illustrate the enhancement of the control performance when adopting the population-games-based dynamical tuning methodology, the performance obtained with the dynamical tuning is compared to the performance when static weights are established to the objectives in the cost function. Besides, two different scenarios are proposed. In general, the water demand profiles have a periodic behavior (daily), with a constant mean value, and with a regular amplitude. Nevertheless, the event in which the periodic demand changes unexpectedly along the time is considered, i.e., when the demand varies its mean value and its regular amplitude. The purpose is to assess the automatic adjustment of the weights when conditions over the system suffer a modification along the time, improving the performance with respect to an MPC with static tuning.

The performance when the demand suffers a decrement, and when demand has a sudden increment are analyzed. These two possible scenarios are presented in Fig. 3.6 where four arbitrarily chosen demands are shown (all the 25 demand profiles for the case study have similar behavior), i.e.,

Scenario 1: decrement of the mean value of the demand profiles (see Fig. 3.6a).
Scenario 2: increment of the mean value of the demand profiles (see Fig. 3.6b).

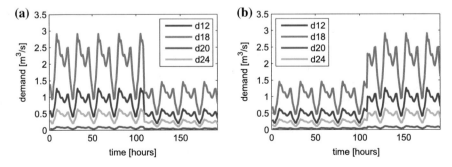

Fig. 3.6 Demand profile for: **a** Scenario 1 and **b** Scenario 2. Disturbances d_{12}, d_{18}, d_{20} and d_{24} correspond to the water demands in the case study presented in Fig. 3.4

The decrement and increment of the mean value of the disturbances is made arbitrarily at the end of the fourth day of simulation.

Notice that when implementing an off-line static tuning strategy, the control designer should assign the desired prioritization weights $\gamma_1, \ldots, \gamma_n$ to the controller. On the other hand, when adopting the dynamical tuning strategy, the control designer assigns the desired prioritization throughout the weights $w_1, \ldots, w_n$. Consequently, in order to make a fair comparison, weights $\gamma_1, \ldots, \gamma_n$ in the cost function of problem (3.4) *for the static tuning case* and the weights in the *management region* $w_1, \ldots, w_n$ in (3.12) *for the dynamical tuning case* are selected to be the same, i.e., $w_i = \gamma_i$, for all $i \in \mathcal{S}$.

3.3.6 Results and Discussion

The performance of the controllers is evaluated by using an economical *Key Performance Index* (KPI) associated to costs denoted by $\mathrm{KPI}_{\mathrm{Ecost}}$, and C considering the total number of simulation days (in this case eight days), i.e.,

$$\mathrm{KPI}_{\mathrm{Ecosts}}(\mathrm{day}) = \sum_{k=1+24(\mathrm{day}-1)}^{24+24(\mathrm{day}-1)} \left(\alpha_1 + \alpha_{2,k}\right)^{\top} \mathbf{u}_k, \qquad (3.15)$$

where $k \in \mathbb{Z}_{\geq 0}$ in given in hours, and

$$C = \sum_{\mathrm{day}=1}^{8} \mathrm{KPI}_{\mathrm{Ecosts}}(\mathrm{day}). \qquad (3.16)$$

Furthermore, a sub-index is used to distinguish between the results with static tuning, and with the proposed dynamical tuning, i.e., C_S and C_D, respectively. For each scenario, six different cases corresponding to six *management regions* are tested, which are outlined as follows:

- Tuning case 1: $[w_1 \quad w_2 \quad w_3]^{\top} = [0.8 \quad 0.05 \quad 0.15]^{\top}$,
- Tuning case 2: $[w_1 \quad w_2 \quad w_3]^{\top} = [0.7 \quad 0.1 \quad 0.2]^{\top}$,
- Tuning case 3: $[w_1 \quad w_2 \quad w_3]^{\top} = [0.6 \quad 0.15 \quad 0.25]^{\top}$,
- Tuning case 4: $[w_1 \quad w_2 \quad w_3]^{\top} = [0.5 \quad 0.2 \quad 0.3]^{\top}$,
- Tuning case 5: $[w_1 \quad w_2 \quad w_3]^{\top} = [0.4 \quad 0.25 \quad 0.35]^{\top}$,
- Tuning case 6: $[w_1 \quad w_2 \quad w_3]^{\top} = [0.35 \quad 0.3 \quad 0.35]^{\top}$,

where $[w_1 \quad w_2 \quad w_3]^{\top} = [\gamma_1 \quad \gamma_2 \quad \gamma_3]^{\top}$. Notice that all the proposed tuning cases satisfy the prioritization order presented in Sect. 3.3.4, i.e., $w_1 > w_3 > w_2$.

Table 3.1 presents the comparison between the economic results obtained with a multi-objective MPC using a static and dynamical population-games-based tuning, and for the two different scenarios. Also, Table 3.1 shows the reduction of costs

Table 3.1 Economic results for Scenario 1 and Scenario 2 in the case study. Notice that for the comparison of data the management region corresponds to the prioritization of the MPC controller with static tuning, i.e., $[w_1 \quad w_2 \quad w_3]^\top = [\gamma_1 \quad \gamma_2 \quad \gamma_3]^\top$

	Tuning case	Dynamical tuning costs C_D (e.u.)	Static tuning costs C_S (e.u.)	Reduction of costs $C_S - C_D$ (e.u.)	Percentage reduction $100(C_S - C_D)/C_S[\%]$
Scenario 1	1	281475.4393	295465.0021	13989.5627	4.73
	2	282296.0113	295114.7771	12818.7657	4.34
	3	283592.6568	300172.7427	16580.0858	5.52
	4	289484.6672	312124.2979	22639.6307	7.25
	5	291048.2900	328267.0268	37218.7368	11.33
	6	291874.0282	341964.3402	50090.3120	14.64
Scenario 2	1	251003.7369	266079.8919	15076.1550	5.66
	2	252147.3533	265056.5038	12909.1505	4.87
	3	255457.0784	270722.0341	15264.9556	5.63
	4	259626.8908	282454.0561	22827.1652	8.08
	5	261713.5740	300459.4927	38745.9187	12.89
	6	263114.2332	313364.1354	50249.9022	16.03

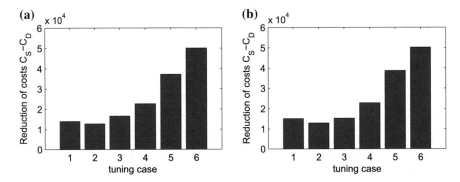

Fig. 3.7 Reduction of costs in eight days for the six different tuning cases. **a** Scenario 1, and **b** Scenario 2

when adopting the proposed dynamical tuning, i.e., $C_S - C_D$. It can be seen that, for all the tested *management regions*, and for both scenarios, a reduction of costs is obtained when implementing a dynamical tuning with respect to the costs with a standard static tuning. Figure 3.7 presents a summary of the reduction of costs for both scenarios and all the tested combination of weights for the control objectives. Cost reductions from 13989.56 to 50090.31 e.u., and from 15076.15 to 50249.90 e.u., are obtained for the first and second scenario in eight days, respectively. Figures 3.8 and 3.9 show the evolution of system states, control inputs, and dynamic prioritization

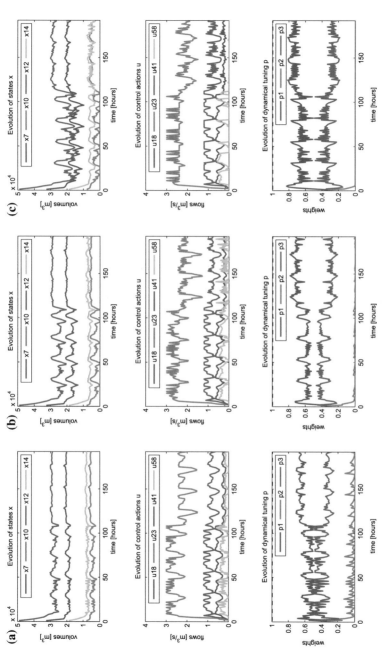

Fig. 3.8 Evolution of volumes, control inputs, and tuning weights for the population-games-based dynamical tuning in Scenario 1, for three different management points: **a** First column $\mathbf{w} = [0.4 \quad 0.25 \quad 0.35]^{\top}$, **b** second column $\mathbf{w} = [0.6 \quad 0.15 \quad 0.25]^{\top}$, and **c** third column $\mathbf{w} = [0.8 \quad 0.05 \quad 0.15]^{\top}$

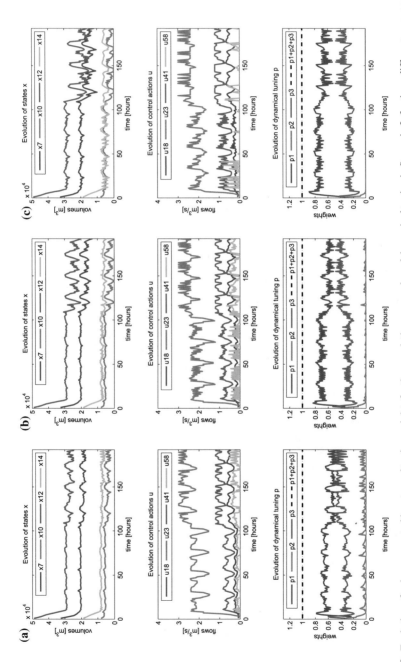

Fig. 3.9 Evolution of volumes, control inputs, and tuning weights for the population-games-based dynamical tuning in Scenario 2, for three different management points: **a** First column $\mathbf{w} = [0.4 \quad 0.25 \quad 0.35]^\top$, **b** second column $\mathbf{w} = [0.6 \quad 0.15 \quad 0.25]^\top$, and **c** third column $\mathbf{w} = [0.8 \quad 0.05 \quad 0.15]^\top$

weights for the first and second scenario with *management regions* given by $\mathbf{w} = [0.4 \quad 0.25 \quad 0.35]^\top$, $\mathbf{w} = [0.6 \quad 0.15 \quad 0.25]^\top$, and $\mathbf{w} = [0.8 \quad 0.05 \quad 0.15]^\top$.

The performance exhibits an oscillatory behavior for the adjustment of weights because of the disturbances in the system. In fact, it can be seen that the periodicity of the oscillation in the weights adjustment corresponds to the diary periodicity of the demands (see Fig. 3.6). In addition, it can be seen in Figs. 3.8 and 3.9 that the dynamical tuning suffers an abrupt change at the end of the fourth day adjusting weights appropriately. This fact occurs since, at that point, the decrement or increment of the mean value for the demand profiles is applied.

3.4 Summary

A novel dynamical tuning methodology for the prioritization weights in multi-objective MPC controllers has been presented in this chapter. The dynamical tuning methodology requires to normalize the cost function of the optimization problem behind the MPC controller. Therefore, a population game is solved with a discrete version of the projection dynamics, which update the appropriate tuning by using information about the current value of the normalized control objectives. The proposed dynamical tuning does not require to generate multiple points of the Pareto front, which implies that it is not computationally costly with respect to other reported on-line approaches. The proposed tuning has been established to be a weighting sum, for which it is required that the sum of all the weights is equal one. It has been shown that the discrete version of the projection dynamics satisfies this constraint throughout the evolution of their variables. Furthermore, the stability analysis of the Nash equilibrium under the discrete projection dynamics has been made, and it is guaranteed if the control objectives decrease as more priority is assigned to them (Assumption 3.1).

Finally, the dynamical tuning methodology is implemented to a large-scale water supply network. Results have shown a reduction of costs when adopting the proposed population-games-based dynamical tuning. The reduction of costs is achieved for all the six tested tuning cases, and for two different scenarios for demand abrupt changes (one scenario considering a decrement of demand, and another considering an increment of demand). It is worth to point out that these achieved cost reductions have been presented for a period of eight days, and that these reductions are maintained along the time. Therefore, the proposed dynamical tuning strategy, according to the results obtained during a week, would represent a bigger reduction of costs in a larger period of time, e.g., a month or a year.

In this chapter, a game theoretical approach has been used to complement an MPC controller. In this regard, game theory helps to enhance the performance of the controller. Similarly, next chapter discusses a different manner in which game theory can collaborate in the design of optimization-based controllers. More specifically, it is explored how a game-theoretical approach can be used for the design of DMPC controllers.

References

1. Grosso J (2015) Economic and robust operation of generalised flow-based networks. Doctoral dissertation. Automatic Control Department. Universidad Politècnica de Catalunya
2. Barreiro-Gomez J, Ocampo-Martinez C, Quijano N (2015) Evolutionary-game-based dynamical tuning for multi-objective model predictive control. In: Olaru S, Grancharova A, Lobo Pereira F (eds) Developments in Model-based Optimization And Control. Springer, Berlin, pp 115–138
3. Barreiro-Gomez J, Ocampo-Martinez C, Quijano N (2017) Dynamical tuning for MPC using population games: a water supply network application. ISA Trans (to appear) 69(2017):175–186
4. Grodzevich O, Romanko O (2006) Normalization and other topics in multi-objective optimization. In: Proceedings of the fields-MITACS industrial problems workshop. Toronto, Ontario, pp 89–101
5. Di Cairano S, Bemporad A (2010) Model predictive control tuning by controller matching. IEEE Trans Autom Control 55:185–190
6. Shah G, Engell S (2011) Tuning MPC for desired closed-loop performance for MIMO systems. In: Proceedings of the American control conference (ACC). San Francisco, USA, pp 4404–4409
7. Waschl H, Alberer D, del Re L (2012) Automatic tuning methods for MPC environments. In: Moreno-Díaz R, Pichler F, Quesada-Arencibia A (eds) Computer aided systems theory - EUROCAST 2011: 13th international conference. Revised selected papers, Part II. Springer, Berlin, pp 41–48. ISBN 978-3-642-27579-1
8. Toro R, Ocampo-Martinez C, Logist F, Van Impe J, Puig V (2011) Tuning of predictive controllers for drinking water networked systems. In: Proceedings of the 18th IFAC world congress. Milan, Italy, pp 14507–14512
9. Grosso JM, Ocampo-Martinez C, Puig V, Joseph B (2014) Chance-constrained model predictive control for drinking water networks. J Process Control 24:504–516
10. Wang Y, Ocampo-Martinez C, Puig V (2016) Stochastic model predictive control based on Gaussian processes applied to drinking water networks. IET Control Theory Appl 10:947–955
11. Wang Y, Ocampo-Martinez C, Puig V, Quevedo J (2016) Gaussian-process-based demand forecasting for predictive control of drinking water networks. In Panayiotou CG, Ellinas G, Kyriakides E, Polycarpou MM (eds) Critical information infrastructures security. Lecture notes in computer science: CRITIS 2014, vol 8985. Springer International Publishing, Berlin, pp 69–80
12. Barreiro-Gomez J, Obando G, Quijano N (2017) Distributed population dynamics: optimization and control applications. IEEE Trans Syst Man Cybern: Syst 47(2):304–314
13. Quijano N, Ocampo-Martinez C, Barreiro-Gomez J, Obando G, Pantoja A, Mojica-Nava E (2017) The role of population games and evolutionary dynamics in distributed control systems. IEEE Control Syst 37(1):70–97
14. Sandholm WH (2010) Population games and evolutionary dynamics. MIT Press, Cambridge
15. Nagurney A, Zhang D (1997) Projected dynamical systems in the formulation, stability analysis, and computation of fixed demand traffic network equilibria. Transp Sci 31:147–158
16. Mesbahi M, Egerstedt M (2010) Graph theoretic methods in multiagent networks. Princeton series in applied mathematics. Princeton University Press, Princeton. ISBN 978-0-691-14061-2
17. Xiao L, Boyd S (2006) Optimal scaling of a gradient method for distributed resource allocation. J Optim Theory Appl 129:469–488
18. Kim IY, de Weck OL (2006) Adaptive weighted sum method for multiobjective optimization: a new method for pareto front generation. Struct Multidiscip Optim 31:105–116
19. Ocampo-Martinez C, Puig V, Cembrano G, Quevedo J (2013) Application of predictive control strategies to the management of complex networks in the urban water cycle. IEEE Control Syst Mag 33(1):15–41
20. Ocampo-Martinez C, Barcelli D, Puig V, Bemporad A (2012) Hierarchical and decentralised model predictive control of drinking water networks: application to Barcelona case study. IET Control Theory Appl 6(1):62–71

Chapter 4
Distributed Predictive Control Using Population Games

In Chap. 3, the role of the population games in the dynamical tuning for MPC controllers has been presented. This chapter presents a different role of population games consisting in the design of DMPC controllers involving resource allocation problems.

The contribution of this chapter is the design of a general method that permits the deduction of several distributed population dynamics, which has been published in [1] and that allows the design of distributed controllers, e.g., an engineering application with a DMPC controller using this distributed population-dynamics approach is presented in [2]. The core of the proposed method is the use of the mean dynamics [3] by including strategy-constrained interactions, extending the results in [4] where the distributed version of one of the six fundamental population dynamics has been presented using a different deduction. For the classical population dynamics approach, the evolution of the proportion of agents depends on the whole population state, i.e., agents can interact with the whole population; however, when adding strategy-constraint interactions, agents can only interact with a portion of the whole population. To illustrate the proposed methodology, a distributed version of the fundamental population dynamics is presented (those obtained by applying classic revision protocols), i.e., the distributed replicator dynamics, the distributed Smith dynamics, the distributed logit dynamics, and the distributed projection dynamics. It is worth noting that the deduction presented in this chapter can be used to generate other distributed dynamics from alternative revision protocols. Besides, it is shown that a population without strategy-constrained interactions obeys a structure given by a complete graph, whereas a population considering strategy-constrained interactions has many different possible structures that are generally given by non-complete graphs. In this sense, the proposed approach is versatile to be implemented in a large variety of problems with different information structures. Moreover, the distributed population dynamics exhibit similar stability and invariance properties as their classic counterpart. Finally, the application of the deduced distributed population dynamics in optimization problems, classic games, and controllers design is highlighted.

© Springer International Publishing AG, part of Springer Nature 2019
J. Barreiro-Gomez, *The Role of Population Games in the Design of Optimization-Based Controllers*, Springer Theses,
https://doi.org/10.1007/978-3-319-92204-1_4

Afterwards an application of the distributed population games to design non-centralized controllers is presented. The formalization of a distributed scheme for a traditional constrained MPC to manage medium/large-scale systems comprised by several sub-systems is discussed. First, the design of a local MPC controller per sub-system that is in charge of managing the desired local variables is proposed. Then, outputs from all controllers are optimally coordinated without the need of a centralized configuration. Different from classical dual decomposition and alternative direction method of multipliers (ADMM), the proposed DMPC scheme based on population dynamics does not require a central coordinator associated to the Lagrange multipliers when managing a coupled constraint involving all the decision variables. Regarding the population dynamics stage in the proposed scheme, under some mild assumptions, it is shown that the solution computed by using the proposed method asymptotically converges to the optimal solution, while all constraints are satisfied. Besides, the stability of the closed-loop system with the DMPC is ensured by proving that there exists an equivalence between the proposed distributed scheme and a centralized MPC controller (CMPC).

4.1   General Dynamics on Graphs

The dynamics describing a population behavior depend on the population with strategy-constrained interactions. In this regard, current literature assumes that the population under consideration is well–mixed, i.e., if any portion of the entire population is taken, this contains all the strategies with the same probability. Figure 4.1a illustrates this fact by showing a population composed by a large and finite number of agents involved in a game. Each element in the figure represents an agent, and the shape of the element (*circle*, *square*, or *triangle*) denotes the strategy that the agent has adopted. In population games, all agents have the same probability to receive a revision opportunity. The agent receiving the revision opportunity randomly chooses another agent from its neighbors and can change its own strategy by the neighbor's

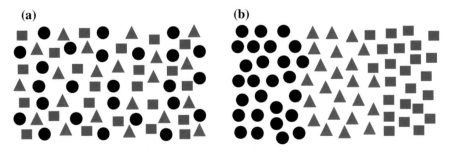

Fig. 4.1 a Well–mixed population, i.e., population without strategy-constrained interactions. **b** Population with strategy-constrained interactions (taken from [1])

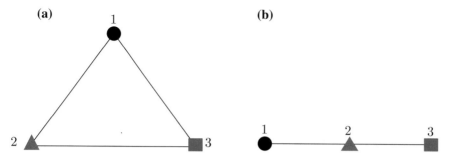

Fig. 4.2 Graph representation of: **a** population without strategy-constrained interactions in Fig. 4.1a, and **b** population with strategy-constrained interactions in Fig. 4.1b (taken from [1])

strategy depending on the selected revision protocol. Since the population is well–mixed and there are no strategy-constrained interactions, the probability that the selected opponent is playing any of the available strategies is the same.

On the other hand, there could be a population considering strategy-constrained interactions as the one shown in Fig. 4.1b. For this population, all agents have the same probability to receive an opportunity to make a revision. However, the probability that the opponent is playing a particular strategy is not equal (e.g., if the strategy played by the agent receiving the revision opportunity is *square*, then there is the same probability to select an opponent playing strategy *triangle* or *square*, but the probability to select an opponent playing strategy *circle* is zero since the population structure does not allow it). Interactions among agents playing different strategies can be represented by a graph $\mathcal{G} = (\mathcal{S}, \mathcal{E}, \mathbf{A})$. The set of nodes $\mathcal{S}$ is associated with the available strategies and the set of links $\mathcal{E}$ is related to the encounter probability between strategies, i.e., there exists a link between two strategies if their encounter probability is different from zero. Hence, the elements of the corresponding adjacency matrix $\mathbf{A} = [a_{ij}]$ are as follows: $a_{ij} = 1$ denotes that strategies i and j can encounter each other, while $a_{ij} = 0$ denotes that the population strategy-constrained interactions make impossible a matching between strategies i and j. According to this convention, the scenarios associated with well–mixed populations and populations with strategy-constrained interactions can be represented by two kinds of graphs. The well–mixed population case is always represented by a complete graph, whereas a population considering strategy-constrained interactions is represented by a graph with a specific topology depending on the particular population structure (see Fig. 4.2). In this chapter, it is assumed that the encounter probability between strategies i and j is the same as the one of strategies j and i, i.e., the graph $\mathcal{G}$ is undirected.

Distributed Mean Dynamics

Taking into account the previously discussed considerations, the evolutionary process of a population with strategy-constrained interactions involved in a strategic game is formally described. Suppose that the population is composed by M agents, and each of them receives a revision opportunity that is given by an exponential distribution with rate R. Hence, during a time dt, the revision opportunity received by each agent is given by Rdt. Since it is assumed that the mass of the population is equal to one, the scalar p_i is equal to the portion of agents playing the ith strategy, and Mp_i is the total number of agents playing strategy $i \in S$. Consequently, the expected number of revision opportunities received by agents playing the ith strategy is approximately $Mp_i Rdt$ during dt (notice that p_i may vary during dt; however, this variation is negligible if dt is small). Agents playing $i \in S$ switch to strategy $j \in S$ with a probability that depends on the revision protocol, the probability distribution of receiving a revision opportunity, and the encounter probability between strategies i and j (given by the population structure, which is represented by the graph $\mathcal{G}$), i.e., $a_{ij}\varrho_{ij}(\mathbf{f}(\mathbf{p}), \mathbf{p})/R$. Finally, the expected number of agents switching from strategy $i \in S$ to strategy $j \in S$ during time dt is $Mp_i a_{ij}\varrho_{ij}(\mathbf{f}(\mathbf{p}), \mathbf{p})dt$.

Now, considering all possible strategies in the population, the expected number of agents switching to strategy $i \in S$ is given by

$$M \sum_{j \in S} p_j a_{ji} \varrho_{ji}(\mathbf{f}(\mathbf{p}), \mathbf{p})dt,$$

and the expected number of agents playing strategy $i \in S$ changing to other strategies is given by

$$Mp_i \sum_{j \in S} a_{ij} \varrho_{ij}(\mathbf{f}(\mathbf{p}), \mathbf{p})dt.$$

Therefore, the variation of the proportion of agents playing the ith strategy is deduced by a mass balance as follows:

$$\dot{p}_i = \sum_{j \in S} p_j a_{ji} \varrho_{ji}(\mathbf{f}(\mathbf{p}), \mathbf{p}) - p_i \sum_{j \in S} a_{ij} \varrho_{ij}(\mathbf{f}(\mathbf{p}), \mathbf{p}), \quad \forall i \in S.$$

This equation corresponds to the *distributed mean dynamics*, or mean dynamics for populations with strategy-constrained interactions. Since $\mathcal{G}$ is undirected then the elements in the adjacency matrix satisfy that $a_{ij} = a_{ji} \in \{0, 1\}$ and $a_{ij} = 0$ if $j \notin \mathcal{N}_i$, notice that the distributed mean dynamics can be rewritten as follows:

$$\dot{p}_i = \sum_{j \in \mathcal{N}_i} p_j \varrho_{ji}(\mathbf{f}(\mathbf{p}), \mathbf{p}) - p_i \sum_{j \in \mathcal{N}_i} \varrho_{ij}(\mathbf{f}(\mathbf{p}), \mathbf{p}), \quad \forall i \in S. \qquad (4.1)$$

For complete graphs, i.e., for well-mixed-populations or populations without strategy-constrained interactions, it is obtained that $\mathcal{N}_i = \mathcal{S}$, getting the classic *Mean Dynamics* [3].

4.2 Distributed Population Dynamics

Distributed mean dynamics allow the inference of population dynamics involving populations with strategy-constrained interactions comprised of agents that are programmed with a specific revision protocol. This section shows the deduction of different distributed population dynamics by using (4.1). The deduced dynamics are named after the classic population dynamics, which are generated by using the corresponding revision protocol.

Distributed Replicator Dynamics (DRD)

The distributed replicator dynamics are obtained from the distributed mean dynamics using the *pairwise proportional imitation* protocol (Table 2.1), as follows:

$$
\begin{aligned}
\dot{p}_i &= \sum_{j \in \mathcal{N}_i} p_j p_i \left[f_i(\mathbf{p}) - f_j(\mathbf{p}) \right]_+ - p_i \sum_{j \in \mathcal{N}_i} p_j \left[f_j(\mathbf{p}) - f_i(\mathbf{p}) \right]_+, \\
&= \sum_{j \in \mathcal{N}_i} p_j p_i (f_i(\mathbf{p}) - f_j(\mathbf{p})), \quad \forall i \in \mathcal{S}.
\end{aligned}
$$

Finally, the distributed replicator dynamics are given by

$$
\dot{p}_i = p_i \left(f_i(\mathbf{p}) \sum_{j \in \mathcal{N}_i} p_j - \sum_{j \in \mathcal{N}_i} p_j f_j(\mathbf{p}) \right), \quad \forall i \in \mathcal{S}. \tag{4.2}
$$

Distributed Smith Dynamics (DSD)

In this case, the *pairwise comparison* protocol is used (see Table 2.1). Substituting this revision protocol in (4.1), it follows:

$$
\dot{p}_i = \sum_{j \in \mathcal{N}_i} p_j \left[f_i(\mathbf{p}) - f_j(\mathbf{p}) \right]_+ - p_i \sum_{j \in \mathcal{N}_i} \left[f_j(\mathbf{p}) - f_i(\mathbf{p}) \right]_+, \quad \forall i \in \mathcal{S}. \tag{4.3}
$$

Notice that (4.3) can be written as

$$\dot{p}_i = \sum_{j \in \mathcal{N}_i} \frac{1}{2}\Big((1 - \nu_{ij})p_i + (1 + \nu_{ij})p_j\Big)\big[f_i(\mathbf{p}) - f_j(\mathbf{p})\big], \quad \forall i \in \mathcal{S},$$

where $\nu_{ij} = \mathrm{sgn}(f_i(\mathbf{p}) - f_j(\mathbf{p}))$.

Distributed Logit Dynamics (DLD)

The deduction of the distributed logit dynamics is based on the *logit choice* protocol (Table 2.1). However, notice that this protocol requires full information since the sum at the denominator is taken over all the strategies. In order to satisfy the information constraint given by the graph $\mathcal{G}$, the protocol is modified as follows:

$$\varrho_{ij}(\mathbf{f}(\mathbf{p}), \mathbf{p}) = e^{\eta^{-1}f_j(\mathbf{p})}, \eta > 0.$$

Distributed logit dynamics are obtained by replacing the above protocol in the distributed mean dynamics, i.e.,

$$\dot{p}_i = \sum_{j \in \mathcal{N}_i} p_j e^{\eta^{-1}f_i(\mathbf{p})} - p_i \sum_{j \in \mathcal{N}_i} e^{\eta^{-1}f_j(\mathbf{p})}, \quad \forall i \in \mathcal{S}. \tag{4.4}$$

Distributed Projection Dynamics (DPD)

The *projection dynamics* use the *modified pairwise comparison* protocol based on the *pairwise comparison* presented in Table 2.1, i.e.,

$$\dot{p}_i = \sum_{j \in \mathcal{N}_i} p_j \frac{[f_i(\mathbf{p}) - f_j(\mathbf{p})]_+}{p_j} - p_i \sum_{j \in \mathcal{N}_i} \frac{[f_j(\mathbf{p}) - f_i(\mathbf{p})]_+}{p_i}, \quad \forall i \in \mathcal{S},$$

$$= \sum_{j \in \mathcal{N}_i} \big(f_i(\mathbf{p}) - f_j(\mathbf{p})\big), \quad \forall i \in \mathcal{S}.$$

Thus, the distributed projection dynamics are given by

$$\dot{p}_i = |\mathcal{N}_i| f_i(\mathbf{p}) - \sum_{j \in \mathcal{N}_i} f_j(\mathbf{p}), \quad \forall i \in \mathcal{S}, \tag{4.5}$$

where $|\mathcal{N}_i|$ denotes the cardinality of the set $\mathcal{N}_i$, i.e., the number of neighbors of the ith node.

4.2.1 Invariant Set Analysis

The population mass does not vary over time. Hence, all possible states generated during the evolution of the population should belong to the simplex Δ given in (2.9). This section shows that the simplex Δ is an invariant set under the distributed population dynamics deduced in the previous section.

Theorem 4.1 *The simplex Δ is an invariant set under: the distributed replicator dynamics (4.2), the distributed Smith dynamics (4.3), and the distributed logit dynamics (4.4).*

Proof According to (2.9), Δ has two conditions, i.e.,

(i) $\sum_{i \in S} p_i = 1$ (mass conservation), and
(ii) $p_i \geq 0$, for all $i \in S$ (non-negativeness).

First, the fact that DRD, DSD, and DLD satisfy condition *(i)* is proven. Notice that this is equivalent to show that $\sum_{i \in S} \dot{p}_i = 0$ under the distributed mean dynamics (4.1). These dynamics can be written by using the adjacency matrix $\mathbf{A}$ of the graph $\mathcal{G}$ as follows:

$$\dot{p}_i = \sum_{j \in S} a_{ij} \varrho_{ji}(\mathbf{f}(\mathbf{p}), \mathbf{p}) p_j - \sum_{j \in S} a_{ij} \varrho_{ij}(\mathbf{f}(\mathbf{p}), \mathbf{p}) p_i, \quad \forall i \in S.$$

Hence,

$$\sum_{i \in V} \dot{p}_i = \sum_{i \in S} \sum_{j \in V} a_{ij} \varrho_{ji}(\mathbf{f}(\mathbf{p}), \mathbf{p}) p_j - \sum_{i \in S} \sum_{j \in S} a_{ij} \varrho_{ij}(\mathbf{f}(\mathbf{p}), \mathbf{p}) p_i.$$

Since $\mathcal{G}$ is undirected (i.e., $a_{ij} = a_{ji}$), then

$$\sum_{i \in S} \dot{p}_i = \sum_{i \in S} \sum_{j \in S} a_{ji} \varrho_{ji}(\mathbf{f}(\mathbf{p}), \mathbf{p}) p_j - \sum_{j \in S} \sum_{i \in S} a_{ji} \varrho_{ji}(\mathbf{f}(\mathbf{p}), \mathbf{p}) p_j$$
$$= 0.$$

Second, it is proven that each dynamic satisfies condition *ii)*:

- DRD: Non-negativeness of each p_i is satisfied given the fact that $\dot{p}_i = 0$ if $p_i = 0$ under distributed replicator dynamics. Thus, if $p_i(0) \geq 0$, then $p_i(t) \geq 0$ for all $t \geq 0$.
- DSD: According to (4.3), notice that when $p_i = 0$ for any $i \in S$, then $\dot{p}_i \geq 0$. Hence, the non-negativeness of p_i is satisfied under distributed Smith dynamics.
- DLD: Notice that $\dot{p}_i \geq 0$ when $p_i = 0$ under distributed Logit dynamics (4.4). Therefore, if $p(0) \in \Delta$, then $p_i(t) \geq 0$ for all $t \geq 0$.

$\square$

Proposition 4.1 *The set* $\Delta' = \left\{ \mathbf{p} \in \mathbb{R}^n : \sum_{i \in \mathcal{S}} p_i = 1 \right\}$ *is invariant under the distributed projection dynamics (4.5).*

Proof The distributed projection dynamics can be written by using the adjacency matrix $\mathbf{A}$ of the graph $\mathcal{G}$ as follows:

$$\dot{p}_i = \sum_{j \in \mathcal{S}} a_{ij} f_i(\mathbf{p}) - \sum_{j \in \mathcal{S}} a_{ij} f_j(\mathbf{p}), \quad \forall i \in \mathcal{S}.$$

Therefore,

$$\sum_{i \in \mathcal{S}} \dot{p}_i = \sum_{i \in \mathcal{S}} \sum_{j \in \mathcal{S}} a_{ij} f_i(\mathbf{p}) - \sum_{i \in \mathcal{S}} \sum_{j \in \mathcal{S}} a_{ij} f_j(\mathbf{p}).$$

Since $a_{ij} = a_{ji}$ because $\mathcal{G}$ is undirected, it follows that

$$\sum_{i \in \mathcal{S}} \dot{p}_i = \sum_{j \in \mathcal{S}} \sum_{i \in \mathcal{S}} a_{ji} f_i(\mathbf{p}) - \sum_{i \in \mathcal{S}} \sum_{j \in \mathcal{S}} a_{ij} f_j(\mathbf{p})$$
$$= 0,$$

which completes the proof.   □

Remark 4.1 It should be noticed that the distributed projection dynamics satisfy one of the conditions of the original simplex Δ, i.e., $\sum_{i \in \mathcal{S}} p_i = 1$ (mass conservation). However, the non-negativeness of p_i is not guaranteed. This fact also occurs in the classic projection dynamics.   ◇

Remark 4.2 Notice that Theorem 4.1 and Proposition 4.1 do not impose any conditions over the interaction graph $\mathcal{G}$. Thus, the studied distributed population dynamics exhibit simplex invariance under any population structure.   ◇

4.2.2 Stability Analysis

Classic population dynamics usually converge to Nash equilibria since they correspond to the expected outcome of games played by rational individuals (i.e., individuals that are trying to maximize their profit). According to the set of Nash equilibria presented in Definition 2.4, in a Nash equilibrium all players perceive the same profit.

This section provides sufficient conditions to guarantee that a Nash equilibrium $\mathbf{p}^\star \in \text{NE}(\mathbf{f})$ of the population game $\mathbf{f}$ is asymptotically stable under the distributed population dynamics derived in Sect. 4.2. These conditions, which are related to the connectivity of the interaction graph and the characteristics of the Nash equilibrium, are summarized in the following assumptions.

Assumption 4.1 The graph $\mathcal{G}$ that describes the population structure is connected.

Assumption 4.2 The Nash equilibrium $\mathbf{p}^{\star} \in \mathrm{NE}(\mathbf{f})$ belongs to the interior of the simplex Δ, i.e., $\mathbf{p}^{\star} \in \mathrm{int}\,\Delta$ as defined in (2.10).

The results on convergence of the distributed population dynamics to a Nash equilibrium are provided below.

Theorem 4.2 *Let* $\mathbf{f}$ *be a full-potential game with strictly-concave potential function* $V(\mathbf{p})$, *and let* $\mathbf{p}^{\star} \in \mathrm{NE}(\mathbf{f})$. *If Assumptions 4.1 and 4.2 hold, then* $\mathbf{p}^{\star}$ *is asymptotically stable under the distributed replicator dynamics (4.2) and the distributed Smith dynamics (4.3).*

Proof Since $\mathbf{p}^{\star} \in \mathrm{NE}(\mathbf{f})$ and $\mathbf{p}^{\star} \in \mathrm{int}\,\Delta$, it is concluded that $f_i(\mathbf{p}^{\star}) = f_j(\mathbf{p}^{\star})$, for all $i, j \in \mathcal{S}$. Moreover, notice that $\mathbf{p}^{\star} = \arg\max_{\mathbf{p} \in \Delta} V(\mathbf{p})$ (applying the Karush–Kuhn–Tucker conditions). Additionally, since $V(\mathbf{p})$ is strictly concave, it is possible to take

$$E(\mathbf{p}) = V(\mathbf{p}^{\star}) - V(\mathbf{p}) \tag{4.6}$$

as a Lyapunov function candidate. The derivative of $E(\mathbf{p})$ along the trajectories of DRD (4.2) and DSD (4.3) is given by

$$\begin{aligned}
\dot{E}(\mathbf{p}) &= -(\nabla V(\mathbf{p}))^{\top}\dot{\mathbf{p}} \\
&= -\mathbf{f}^{\top}\dot{\mathbf{p}} \\
&= -\mathbf{f}^{\top}\mathbf{L}^{(\mathrm{p})}\mathbf{f},
\end{aligned}$$

where $\mathbf{L}^{(\mathrm{p})} = \left[l_{ij}^{(\mathrm{p})}\right]$ is a matrix whose entries $l_{ij}^{(\mathrm{p})}$ are for DRD as follows:

$$l_{ij}^{(\mathrm{p})} = \begin{cases} -a_{ij}\,p_i\,p_j, & \text{if } i \neq j \\ \displaystyle\sum_{k \in \mathcal{S}, k \neq i} a_{ik}\,p_i\,p_k, & \text{if } i = j, \end{cases}$$

and for DSD as follows:

$$l_{ij}^{(\mathrm{p})} = \begin{cases} -\dfrac{a_{ij}}{2}\Big((1 - \phi_{ij})p_i + (1 + \phi_{ij})p_j\Big), & \text{if } i \neq j, \\ \displaystyle\sum_{k \in \mathcal{S}, k \neq i} \dfrac{a_{ik}}{2}\Big((1 - \phi_{ik})p_i + (1 + \phi_{ik})p_k\Big), & \text{if } i = j. \end{cases}$$

Notice that $\mathbf{L}^{(\mathrm{p})}$ is the Laplacian of the undirected graph given by the tuple $\mathcal{G}^{(\mathrm{p})} = (\mathcal{S}, \mathcal{E}, \mathbf{A}^{(\mathrm{p})})$, where $\mathbf{A}^{(\mathrm{p})} = [a_{ij}^{(\mathrm{p})}]$ is the adjacency matrix whose entries are defined as follows:

$$a_{ij}^{(\mathrm{p})} = \begin{cases} a_{ij}\,p_i\,p_j, & \text{for DRD,} \\ \dfrac{a_{ij}}{2}\Big((1 - \phi_{ij})p_i + (1 + \phi_{ij})p_j\Big), & \text{for DSD.} \end{cases}$$

These entries are nonnegative since $\mathbf{p} \in \Delta$. Thus, $\mathbf{L}^{(\mathbf{p})} \geq 0$ and $\dot{E}(\mathbf{p}) \leq 0$. Therefore, $\mathbf{p}^{\star}$ is stable under DRD and DSD.

Considering that $\mathbf{p}^{\star} \in \text{int}\Delta$ is stable, a set $\mathcal{B}$ around $\mathbf{p}^{\star}$ can be defined such that if $\mathbf{p}(0) \in \mathcal{B}$, then $\mathbf{p}(t) \in \text{int}\Delta$, for all $t \geq 0$ (it is possible to show that $\mathcal{B} = \text{int}\Delta$ for DRD). Thus, if $\mathbf{p}(0) \in \mathcal{B}$, the null space of $\mathbf{L}^{(\mathbf{p})}$ is equal to $\text{span}\{\mathbb{1}_n\}$ since $\mathcal{G}^{(\mathbf{p})}$ is connected. It is concluded that $\mathcal{G}^{(\mathbf{p})}$ is connected since:

- $\mathcal{G}^{(\mathbf{p})}$ and $\mathcal{G}$ have the same topology in $\mathcal{B}$, i.e., if $\mathbf{p} \in \mathcal{B}$, $a_{ij}^{(\mathbf{p})} = 0$ only if $a_{ij} = 0$,
- $\mathcal{G}$ is connected by assumption.

In this case, $\dot{E}(\mathbf{p}) = 0$ if and only if $f_i = f_j$, for all $i, j \in S$, i.e., $\dot{E}(\mathbf{p}) = 0$ only in $\mathbf{p}^{\star}$. Therefore, $\mathbf{p}^{\star}$ is asymptotically stable.   □

Remark 4.3 Theorem 4.2 requires that, in steady state, all strategies are played by the individuals involved in the game. Indeed, when any proportion of individuals is extinct at equilibrium (i.e., $p_i^{\star} = 0$ for some $i \in S$), then convergence of the distributed replicator equation and the distributed Smith dynamics to the Nash equilibrium is not guaranteed. However, the same arguments used in [4] can be employed to relax the convergence conditions. In fact, if subtracting from the original graph any set of nodes associated to extinct strategies does not produce disconnected subgraphs, then convergence to Nash equilibria is provable even if Nash equilibria do not belong to the interior of Δ. This relaxed assumption often holds in well-connected graphs.[1]   ◇

Remark 4.4 Notice that Theorem 4.2 is only applicable to full-potential games. However, this class of games arises in a large number of applications including resource allocation problems and congestion games [3].   ◇

Once the stability analysis for the DRD and the DSD have been presented, the stability of the equilibrium $\mathbf{p}^{\star} \in \text{NE}(\mathbf{f})$ for the DPD is presented in Theorem 4.3.

Theorem 4.3 *Let $\mathbf{f}$ be a continuously differentiable stable game, let $\mathbf{p}^{\star} \in \text{NE}(\mathbf{f})$, and let $\dot{\mathbf{p}}$ be the distributed projection dynamics (4.5). If Assumptions 4.1 and 4.2 hold, then $\mathbf{p}^{\star}$ is asymptotically stable.*

Proof Consider the pairwise comparison protocol $\varrho_{ij} = [f_j(\mathbf{p}) - f_i(\mathbf{p})]_+$, and define $\varrho_{ij} = \varphi(f_j(\mathbf{p}) - f_i(\mathbf{p}))$, where $\varphi(\cdot) = [\cdot]_+$ (where $\max(0, \cdot) = [\cdot]_+$ as in Table 2.1). Then, consider the Lyapunov function candidate:

$$V(\mathbf{p}) = \sum_{i \in S} \sum_{j \in S} a_{ij} \int_0^{f_j(\mathbf{p}) - f_i(\mathbf{p})} \varphi(s)ds.$$

---

[1] A graph is considered to be well connected if the connectivity of the graph does not depend on *few* nodes, e.g., a path graph is not well connected since the removal of any node involving two edges would disconnect the graph (connectivity relies on $n - 2$ nodes), whereas a complete graph is considered well connected since the removal of a node does not imply the disconnection of the graph (connectivity does not depend on any node).

Since $\varphi : \mathbb{R} \to \mathbb{R}_{\geq 0}$ is increasing on $[0, +\infty)$ and $\mathcal{G}$ is connected, then the function $V(\mathbf{p}) > 0$, for all $\mathbf{p} \neq \mathbf{p}^\star$. Additionally, $V(\mathbf{p}^\star) = 0$ since $f_j(\mathbf{p}^\star) = f_i(\mathbf{p}^\star)$, for all $i, j \in \mathcal{S}$. Moreover, notice that

$$\frac{\partial V(\mathbf{p})}{\partial p_\ell} = \sum_{i \in \mathcal{S}} \sum_{j \in \mathcal{S}} a_{ij} \left(\frac{\partial f_j(\mathbf{p})}{\partial p_\ell} - \frac{\partial f_i(\mathbf{p})}{\partial p_\ell} \right) \varphi(f_j(\mathbf{p}) - f_i(\mathbf{p}))$$

$$= \sum_{i \in \mathcal{S}} \sum_{j \in \mathcal{S}} a_{ij} \varphi(f_j(\mathbf{p}) - f_i(\mathbf{p})) \frac{\partial f_j(\mathbf{p})}{\partial p_\ell} - \sum_{j \in \mathcal{S}} \sum_{i \in \mathcal{S}} a_{ji} \varphi(f_i(\mathbf{p}) - f_j(\mathbf{p})) \frac{\partial f_j(\mathbf{p})}{\partial p_\ell}.$$

Taking into account that $a_{ij} = a_{ji}$, it yields

$$\frac{\partial V(\mathbf{p})}{\partial p_\ell} = \sum_{i \in \mathcal{S}} \sum_{j \in \mathcal{S}} a_{ji} (f_j(\mathbf{p}) - f_i(\mathbf{p})) \frac{\partial f_j(\mathbf{p})}{\partial p_\ell}$$

$$= \sum_{j \in \mathcal{S}} \frac{\partial f_j(\mathbf{p})}{\partial p_\ell} \sum_{i \in \mathcal{N}_j} (f_j(\mathbf{p}) - f_i(\mathbf{p})).$$

According to (4.5), notice that $\sum_{i \in \mathcal{N}_j} (f_j(\mathbf{p}) - f_i(\mathbf{p})) = \dot{p}_j$, where $\dot{p}_j$ is the jth element of the distributed projection dynamics $\dot{\mathbf{p}}$. Hence,

$$\frac{\partial V(\mathbf{p})}{\partial p_\ell} = \sum_{j \in \mathcal{S}} \dot{p}_j \frac{\partial f_j(\mathbf{p})}{\partial p_\ell}. \tag{4.7}$$

Therefore, the time derivative of the Lyapunov function is

$$\dot{V}(\mathbf{p}) = (\nabla V(\mathbf{p}))^\top \dot{\mathbf{p}}$$
$$= \dot{\mathbf{p}}^\top D\mathbf{f}(\mathbf{p}) \dot{\mathbf{p}},$$

where $\dot{\mathbf{p}}^\top D\mathbf{f}(\mathbf{p}) \dot{\mathbf{p}} \leq 0$ since $\mathbf{f}$ is stable. □

Remark 4.5 As was stated in the proof of Theorem 4.2, the fact that a Nash equilibrium $\mathbf{p}^\star$ belongs to $\mathrm{int}\,\Delta$ implies that all the fitness functions reach the same value, i.e., $f_i(\mathbf{p}^\star) = f_j(\mathbf{p}^\star)$, for all $i, j \in \mathcal{S}$. Therefore, the results given in Theorems 4.2 and 4.3 are related to the contributions reported in the literature on consensus in multi-agent networks (e.g., see [5–7]).[2] An essential difference is that Theorems 4.2 and 4.3 show a direct relationship between game-theoretic properties and Lyapunov stability of a population game under distributed dynamics. ◇

The connectivity condition of the graph $\mathcal{G}$ in Theorems 4.2 and 4.3 is sufficient for Nash equilibrium stability. Regarding this fact, it is interesting to study if this condition is also necessary. The following proposition gives insights on this issue.

---

[2]In the consensus problem, it is desired to achieve the agreement among different variables, e.g., $p_i^\star = p_j^\star$, for all $i, j \in \{1, \ldots, n\}$, objective that can be achieved taking advantage of the convergence result when $\mathbf{p}^\star \in \mathrm{int}\,\Delta$, i.e., $f_i(\mathbf{p}^\star) = f_j(\mathbf{p}^\star)$ under the population-game framework.

Proposition 4.2 *Assume that the population game* **f** *has a unique Nash equilibrium, which is in the interior of the simplex* Δ, *i.e.,* $\mathbf{p}^\star \in \mathrm{int}\,\Delta$. *Let* $\dot{\mathbf{p}}$ *be the distributed mean dynamics (4.1). If* $\mathbf{p}(t)$ *asymptotically converges to* $\mathbf{p}^\star$, *for all* $\mathbf{p}(0) \in \mathrm{int}\,\Delta$, *then the graph* $\mathcal{G}$ *is connected.*

Proof The proof is developed showing the contrapositive. Assume that $\mathcal{G}$ is non-connected. The graph $\mathcal{G}$ can be expressed as the union of $r \geq 2$ connected components (maximal connected sub-graphs) denoted by $\mathcal{G}^\ell = (\mathcal{S}^\ell, \mathcal{E}^\ell)$, where $\ell = 1, \ldots, r$, i.e., $\mathcal{G} = \bigcup_{\ell=1}^{r} \mathcal{G}^\ell$. Using the arguments in the proof of Theorem 4.1, it can be concluded that, under the distributed mean dynamics, $\sum_{i \in \mathcal{S}^\ell} p_i(t) = \sum_{i \in \mathcal{S}^\ell} p_i(0)$, for all $\ell = 1, \ldots, r$, and for all $t \geq 0$. Take two connected components $\mathcal{G}^1$ and $\mathcal{G}^2$ of the graph $\mathcal{G}$. Furthermore, take the following initial condition:

$$
p_i(0) = \begin{cases} p_i^\star + \frac{\varepsilon}{|\mathcal{S}^1|} & \text{if} \quad i \in \mathcal{S}^1, \\ p_i^\star - \frac{\varepsilon}{|\mathcal{S}^2|} & \text{if} \quad i \in \mathcal{S}^2, \\ p_i^\star & \text{otherwise,} \end{cases}
$$

where $i \in \mathcal{S}$, and $\varepsilon > 0$. Notice that, for small values of ε, $\mathbf{p}(0) \in \mathrm{int}\,\Delta$ since $\mathbf{p}^\star \in \mathrm{int}\,\Delta$. Under this initial condition, it is not possible that $\mathbf{p}(t)$ converges to the unique Nash equilibrium $\mathbf{p}^\star$ since $\sum_{i \in \mathcal{S}^1} p_i(t) = \sum_{i \in \mathcal{S}^1} p_i(0) > \sum_{i \in \mathcal{S}^1} p_i^\star$, for all $t \geq 0$. $\qquad\square$

Therefore, connectivity of $\mathcal{G}$ is required to guarantee convergence to the Nash equilibrium from any initial condition inside the simplex Δ. However, this condition might not be necessary if the initial conditions $\mathbf{p}(0)$ are constrained satisfying an extra condition as it is explained next. For instance, suppose that the graph $\mathcal{G}$ in the population game is non-connected. Moreover, $\mathcal{G}$ is composed of r connected components (maximal connected sub-graphs) denoted by $\mathcal{G}^\ell = (\mathcal{V}^\ell, \mathcal{E}^\ell)$, where $\ell = 1, \ldots, r$, i.e., $\mathcal{G} = \bigcup_{\ell=1}^{r} \mathcal{G}^\ell$. Then, it can be shown (following the same reasoning as in proof of Proposition 4.2) that the equilibrium point $\mathbf{p}^\star \in \mathrm{int}\,\Delta$ is asymptotically stable if $\sum_{i \in \mathcal{S}^\ell} p_i^\star = \sum_{i \in \mathcal{S}^\ell} p_i(0)$, for all $\ell = 1, \ldots, r$. Consequently, in this case, the connectivity condition of $\mathcal{G}$ is not necessary.

4.2.3 Solving Distributed Constrained Optimization Problems

First, the following distributed optimization problem is proposed as an illustrative example

$$
\text{maximize} \quad V(\mathbf{p}) := -\mathbf{p}^\top \mathbf{p} + \mathbf{b}^\top \mathbf{p}, \tag{4.8a}
$$

$$
\text{subject to} \quad \sum_{i=1}^{50} p_i = 1, \tag{4.8b}
$$

$$
p_i > 0, \quad \forall i = 1, \ldots, 50, \tag{4.8c}
$$

Fig. 4.3 Non–complete
graph for a distributed
optimization illustrative
example (taken from [1])

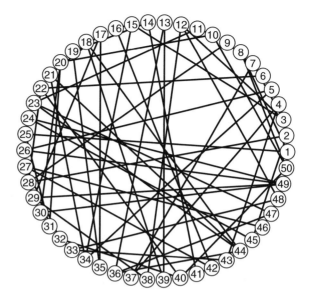

where $\mathbf{p} \in \mathbb{R}_{\geq 0}^{50}$ is the vector of decision variables, and $\mathbf{b} \in \mathbb{R}^{50}$ is a constant vector
of constants, whose entries are given by $b_i = \frac{2i}{1275}$, i.e., $\mathbf{b} = \frac{1}{1275}[2 \quad 4 \quad \ldots \quad 100]^\top$.
Each decision variable is managed by a node in a network. Furthermore, information
constraints given by the graph shown in Fig. 4.3 are imposed. This graph is obtained
by following the Erdös–Rényi model (which is the simplest model of several kind of
social and biological networks [8]) with edge generation probability equal to 0.01.
Besides, a path connecting all nodes is added to guarantee that the generated graph is
connected. The information constraint implies that the ith node only has information
about the state of its neighbors.

In order to solve the problem in (4.8), a full-potential game $\mathbf{f}(\mathbf{p}) = \left[\frac{\partial V}{\partial p_1} \quad \ldots \quad \frac{\partial V}{\partial p_{50}} \right]^\top$ is defined (i.e., the fitness functions correspond to the marginal
utilities) and the distributed population dynamics derived in Sect. 4.2 are applied.
Notice that all nodes satisfy the information constraints (this fact is not possible by
using the classic population dynamics). Results are shown in Fig. 4.4 considering an
initial condition $p_i(0) = \frac{1}{50}$, for all $i = 1, \ldots, 50$. The first and fourth row of Fig. 4.4
show that $\mathbf{p}(t)$ satisfies the problem constraints for all time, i.e., $p_i(t)$ remains non-
negative, for all $i = 1, \ldots, 50$; and $\sum_{i=1}^{50} p_i(t) = 1$. Furthermore, the third row of
Fig. 4.4 shows that all distributed dynamics increase the objective function $V(\mathbf{p})$.
However, only DRD, DSD, and DPD reach the optimum value (which are depicted
in dashed-red line). According to the second row of Fig. 4.4, DRD, DSD, and DPD
equalize the values of fitness functions in steady state, i.e., these dynamics converge
to a Nash equilibrium.

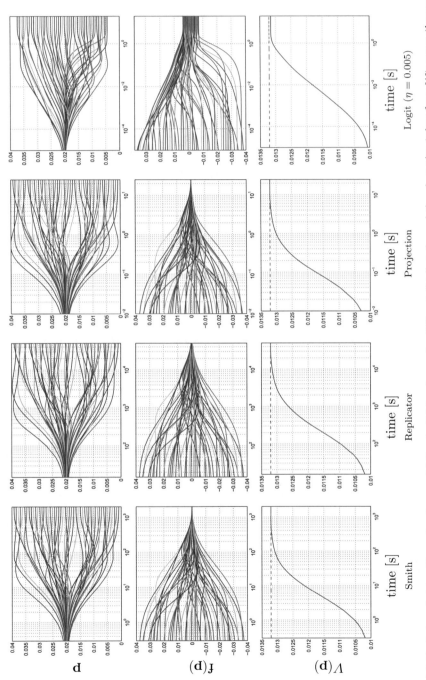

Fig. 4.4 Evolution of states, fitness functions and full-potential function under different distributed population dynamics (taken from [1]): states (1st row), fitness functions (2nd row), objective function (3rd row)

This behavior is consistent with the results stated in Theorems 4.2 and 4.3 since $V(\mathbf{p})$ corresponds to a strictly concave potential function, i.e., $\mathbf{f}$ is a full-potential and stable game. Moreover, convergence time varies from one dynamic to another. DLD show the fastest time response while the convergence of DRD is the slowest.

4.3 DMPC Design

In this section, a DMPC controller is designed by using the distributed population dynamics presented in Sect. 4.2 (in this regard, the distributed population dynamics collaborate in the design of the DMPC taking advantage of their distributed features and stability properties). Due to the fact that there are already many existing approaches in the design of this type of controllers, some differences are pointed out with respect to some methods found in the literature.

DMPC has been extensively studied in recent years (see [9] for a comprehensive review). Two main groups of DMPC techniques are distinguished: those that require a centralized coordinator (e.g., ADMM based techniques [10]), and those that are fully distributed. The proposed approach in this chapter falls in the second group, which is more suitable for applying in problems involving low-bandwidth communications. Among fully distributed methods, there are a variety of approaches. One of the most employed is the one based on distributed ADMM (D-ADMM), which is a version of the ADMM technique that eliminates the need of a centralized coordinator. This approach was first proposed in [11] and it is possible to find several extensions in the literature (e.g., [12]). Although D-ADMM-based techniques are flexible and efficient, they require more complex communications for solving the problem addressed in this section than the communications needed by the proposed approach with population games. This advantage of the method proposed in this section comes from the fact that each node in D-ADMM handles a set of manipulated variables while in the proposed approach introduced in this chapter, each node handles only a single variable. Therefore, the computational operations required by each node are also simplified. Other methods based on cooperative control have also been proposed in the literature to address DMPC problems, e.g., [13]. Cooperative-control-based methods deal with dynamically-coupled plants. However, they generally do not consider coupled constraints on control inputs as is the case of the formulation developed in this section. Some extensions of cooperative DMPC capable to deal with coupled constraints have been reported. However, they either require that each agent has complete information of the constraints [14] or need a centralized agent at some step of the algorithm [15]. Regarding the plug-and-play features in non-centralized MPC controllers, the main challenge consists in reducing the amount of required modifications over the design when plugging in or unplugging sub-systems from the whole system. For instance, in [16], when sub-systems are added or removed, the distributed plug-and-play solution requires to redesign a certain number of controllers. Additionally, in this chapter it is assumed that sub-systems get plugged in and unplugged in an off-line manner. In [17], before the modification is made by plugging in or unplugging sub-systems, the feasibility should be verified. Moreover, when a new sub-system i is connected to a sub-system j, then the jth controller should be redesigned. Likewise, the plug-and-play feature presented in [18] implies the modification of local control laws when it is desired to plug in or unplug sub-systems. Also, it is assumed that the modification is notified previously to the corresponding neighborhood by a request message. Differently, the advantage of the plug-and-play feature of the proposed method in this chapter is that it does not require to modify any of the existing local controllers. In addition, the plug-and-play can be performed in an on-line manner.

4.3.1   Control Problem Statement

Consider a large-scale system composed by m controllable sub-systems that are connected by a communication network. The topology of the communication network is given by an undirected and connected graph denoted by $\tilde{\mathcal{G}} = (\tilde{\mathcal{S}}, \tilde{\mathcal{E}}, \tilde{\mathbf{A}})$. Let $\tilde{\mathcal{S}}$ be the set of nodes that represents the m sub-systems, $\tilde{\mathcal{E}} = \{(i, j) : i, j \in \tilde{\mathcal{S}}\}$ the set of links representing the available communication and/or information sharing among sub-systems, $\tilde{\mathbf{A}}$ is the adjacency matrix, and $\tilde{\mathcal{N}}_i = \{j : (i, j) \in \tilde{\mathcal{E}}\}$ is the set of neighbors of the node $i \in \tilde{\mathcal{S}}$. Notice that in general, the graph $\mathcal{G} \neq \tilde{\mathcal{G}}$, since $\tilde{\mathcal{G}}$ defines the graph representing the system structure of m sub-systems, whereas $\mathcal{G}$ represents the graph topology for all the control inputs within the system (see for example Fig. 4.5).

Each controllable sub-system has a linear time-invariant discrete dynamics given by

$$\mathbf{x}_{i,k+1} = \mathbf{A}_{d,i}\mathbf{x}_{i,k} + \mathbf{B}_{d,i}\mathbf{u}_{i,k}, \tag{4.9}$$

where $k \in \mathbb{Z}_{\geq 0}$ denotes the discrete time step, $i \in \tilde{\mathcal{S}} = \{1, \ldots, m\}$ is the sub-system index, $\mathbf{x}_i \in \mathbb{R}^{n_{x,i}}$ denotes the system states vector, $\mathbf{u}_i \in \mathbb{R}^{n_{u,i}}$ denotes the control inputs vector of the ith sub-system, and matrices $\mathbf{A}_{d,i} \in \mathbb{R}^{n_{x,i} \times n_{x,i}}$ and $\mathbf{B}_{d,i} \in \mathbb{R}^{n_{x,i} \times n_{u,i}}$ have constant elements. The optimization problem behind the MPC controller, denoted by $\mathcal{P}_{\text{MPC}}$, can be stated as follows:

Fig. 4.5 Example of three sub-systems $m = 3$, where $\mathbf{u}_1 \in \mathbb{R}^3$, $\mathbf{u}_2 \in \mathbb{R}^2$, and $\mathbf{u}_3 \in \mathbb{R}^5$, i.e., $n_{u,1} = 3$, $n_{u,2} = 2$, and $n_{u,3} = 5$. Then, $n = \sum_{j=1}^{m} n_{u,j} = 10$

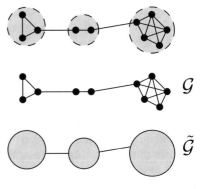

$$\underset{\mathbf{u}_{i,k|k},\ldots,\mathbf{u}_{i,k+H_p-1|k},\ \forall i=1,\ldots,m}{\text{minimize}} \quad J_k = \sum_{i=1}^{m} \left\{ J_i^f \left(\mathbf{x}_{i,k+H_p|k} \right) + \sum_{j=0}^{H_p-1} J_i^\ell \left(\mathbf{x}_{i,k+j|k}, \mathbf{u}_{i,k+j|k} \right) \right\},$$

(4.10a)

subject to

$$\mathbf{x}_{i,k+1+j|k} = \mathbf{A}_{d,i}\mathbf{x}_{i,k+j|k} + \mathbf{B}_{d,i}\mathbf{u}_{i,k+j|k}, \ \forall i \in \tilde{\mathcal{S}}, j \in [0, H_p] \cap \mathbb{Z}_{\geq 0},$$

(4.10b)

$$\mathbf{x}_{i,k+j|k} \in \mathcal{X}_i, \ \forall i \in \tilde{\mathcal{S}}, j \in [0, H_p] \cap \mathbb{Z}_{\geq 0},$$ (4.10c)

$$\mathbf{u}_{i,k+j|k} \in \mathcal{U}_i, \ \forall i \in \tilde{\mathcal{S}}, j \in [0, H_p - 1] \cap \mathbb{Z}_{\geq 0}$$ (4.10d)

$$\sum_{i=1}^{m} \mathbf{1}_{n_{u,i}}^\top \mathbf{u}_{i,k+j|k} \leq \pi, \ \forall j \in [0, H_p - 1] \cap \mathbb{Z}_{\geq 0}$$ (4.10e)

The sets $\mathcal{X}_i \triangleq \{\mathbf{x}_i \in \mathbb{R}^{n_{x,i}} : \underline{\mathbf{x}}_i \leq \mathbf{x}_i \leq \bar{\mathbf{x}}_i\}$, and $\mathcal{U}_i \triangleq \{\mathbf{u}_i \in \mathbb{R}^{n_{u,i}} : \underline{\mathbf{u}}_i \leq \mathbf{u}_i \leq \bar{\mathbf{u}}_i\}$. Moreover, vectors $\underline{\mathbf{x}}_i$ and $\bar{\mathbf{x}}_i$ determine the minimum and maximum state bounds of the ith sub-system, respectively, and $\underline{\mathbf{u}}_i$ and $\bar{\mathbf{u}}_i$ determine the minimum and maximum control input bounds, respectively. The value of $\pi \in \mathbb{R}_{>0}$ determines the total available resource as an *energy* constraint for the whole system. The cost function J_k in (4.10) penalizes the state error and the magnitude of the control inputs for all sub-systems over the prediction horizon $H_p \in \mathbb{Z}_{>0}$. The cost function at time instant k to minimize in the optimization problem behind the MPC controller is denoted by J_k. Moreover the cost function throughout the prediction horizon for the ith sub-system is denoted by J_i^ℓ, whereas the terminal cost corresponding to the ith sub-system is denoted by J_i^f. The terms in the cost function J_k of the optimization problem (4.10) are of the following form [19, 20]:

$$J_i^\ell \left(\mathbf{x}_{i,k}, \mathbf{u}_{i,k}, \mathbf{r}_{i,k} \right) = \|\mathbf{x}_{i,k} - \mathbf{r}_{i,k}\|_{\mathbf{Q}_i}^2 + \|\mathbf{u}_{i,k}\|_{\mathbf{R}_i}^2,$$

$$J_i^f \left(\mathbf{x}_i \right) = \begin{cases} J_i^c \left(\mathbf{x}_i \right), & \text{if } \mathbf{x}_i \in \mathcal{X}_i^f, \\ \pi_i^c, & \text{if } \mathbf{x}_i \notin \mathcal{X}_i^f, \end{cases}$$ (4.11)

where $\mathbf{Q}_i \in \mathbb{R}^{n_{x,i} \times n_{x,i}}$ is a positive semi-definite weighting matrix related to the system states, and $\mathbf{R}_i \in \mathbb{R}^{n_{u,i} \times n_{u,i}}$ is a positive definite weighting matrix related to the control inputs. Vector $\mathbf{r}_{i,k}$ is the reference for the ith sub-system. The function J_i^c is continuous and $J_i^c (\mathbf{x}_i) \geq 0$, for all $\mathbf{x}_i \in \mathcal{X}_i$, and the set $\mathcal{X}_i^f = \{\mathbf{x}_i \in \mathcal{X}_i : J_i^c (\mathbf{x}_i) \leq \pi_i^c\}$. Moreover, the steady state of the system is given by $\mathbf{x}_i^s \in \mathcal{X}_i^f$. Consequently, J_i^f is also continuous. For instance, one selection may be $J_i^c (\mathbf{x}_k) = \|\mathbf{x}_{i,k} - \mathbf{r}_{i,k}\|_{\mathbf{V}_i}^2$, where $\mathbf{V}_i \in \mathbb{R}^{n_{x,i} \times n_{x,i}}$ is a positive semi-definite weighting matrix related to the state error at the prediction horizon.

Notice that if constraints depending on the reference signals are added to the optimization problem in (4.10), then the terminal cost in (4.11) can be changed, and some claims about the steady state $\mathbf{x}_i^s$ can be made. There are two possible options in order to incorporate these type of constraints:

- Consider the function $J_i^c (\mathbf{x}_k) = \|\mathbf{x}_{i,k} - \mathbf{r}_{i,k}\|_{\mathbf{V}_i}^2$, and a terminal constraint given by $\|\mathbf{x}_{i,k+H_p|k} - \mathbf{r}_{i,k+H_p|k}\|_{\mathbf{V}_i}^2 \leq \pi_i^c$. In this case, if the problem (4.10) is feasible, then $\|\mathbf{x}_i^s - \mathbf{r}_{i,k+H_p|k}\|_{\mathbf{V}_i}^2 \leq \pi_i^c$ and the terminal cost is $J_i^f (\mathbf{x}_{i,k+H_p|k}) = \|\mathbf{x}_{i,k+H_p|k} - \mathbf{r}_{i,k+H_p|k}\|_{\mathbf{V}_i}^2$, which corresponds to an MPC controller with terminal-set constraint.
- Considering the function $J_i^c (\mathbf{x}_k) = \|\mathbf{x}_{i,k} - \mathbf{r}_{i,k}\|_{\mathbf{V}_i}^2$ and a terminal constraint given by $\|\mathbf{x}_{i,k+H_p|k} - \mathbf{r}_{i,k+H_p|k}\|_{\mathbf{V}_i}^2 = 0$, then if the problem (4.10) is feasible, it follows that $\mathbf{x}_i^s = \mathbf{r}_{i,k+H_p|k}$, which corresponds to an MPC controller with terminal-point constraint.

The control sequence along H_p is denoted as in (3.3a)

$$\hat{\mathbf{u}}_k \triangleq \left(\mathbf{u}_{k|k}, \mathbf{u}_{k+1|k}, \ldots, \mathbf{u}_{k+H_p-1|k}\right),$$

and if the optimization problem is feasible, there exists an optimal control sequence

$$\hat{\mathbf{u}}_k^\star \triangleq \left(\mathbf{u}_{k|k}^\star, \mathbf{u}_{k+1|k}^\star, \ldots, \mathbf{u}_{k+H_p-1|k}^\star\right)$$

that minimizes the cost J_k and that generates an optimal trajectory as in (3.3b)

$$\hat{\mathbf{x}}_k \triangleq \left(\mathbf{x}_{k+1|k}, \mathbf{x}_{k+2|k}, \ldots, \mathbf{x}_{k+H_p|k}\right).$$

The optimal control input that is applied to the system is denoted by $\mathbf{u}_{\mathrm{MPC},k} = \mathbf{u}_{k|k}^\star$.

Prior to presenting the stability analysis of the population-games-based DMPC controller, some assumptions are considered in order to define the framework of the proposed scheme. Therefore, it is assume that there exist values for $J_i^f : \mathcal{X}_i \to \mathbb{R}$, and π_i^c such that assumptions are hold. First, consider the basic stability assumption stated in [21].

Assumption 4.3 For each state vector $\mathbf{x}_{i,k} \in \mathcal{X}_i^f$, there exists a control input $\mathbf{u}_i \in \mathcal{U}_i$ such that $\mathbf{x}_{i,k+1} \in \mathcal{X}_i^f$ under the system (4.9), i.e,

$$\underset{\mathbf{u}_{i,k} \in \mathcal{U}_i}{\text{minimize}} \left\{ J_i^f (\mathbf{x}_{i,k+1}) + J_i^\ell (\mathbf{x}_{i,k}, \mathbf{u}_{i,k}) : \mathbf{x}_{i,k+1} \in \mathcal{X}_i^f \right\} \leq J_i^f (\mathbf{x}_{i,k}), \qquad (4.12)$$

where (4.12) holds for all $\mathbf{x}_{i,k} \in \mathcal{X}_i^f$.   ◇

Besides, according to [21] Assumption 4.3 implies the invariance of the set $\mathcal{X}_i^f$ in Assumption 4.4.

Assumption 4.4 The set $\mathcal{X}_i^f$ is control invariant for the system (4.9). Consequently, if $\mathbf{x}_{i,k} \in \mathcal{X}_i^f$, then $\mathbf{x}_{i,k+1} \in \mathcal{X}_i^f$.   ◇

Assumption 4.5 It is assumed that, in case there is not enough resource to achieve the established set-points, the system in (4.9) admits a steady state with the model (4.10b) and constraints (4.10c)–(4.10e).   ◇

Notice that if $\mathbf{A}_{d,i}$ in (4.9) does not have eigenvalues at unity, then any control input $\mathbf{u}_{i,k} \in \mathcal{U}$ admits a steady state. However, Assumption 4.5 is stated since if $\mathbf{A}_{d,i}$ has an unitary eigenvalue then the system (4.9) and constraints (4.10c)–(4.10e) might not admit a steady state [22].

Furthermore, due to the fact that the optimization problem (4.10) defines a limited resource problem, it is possible that a sub-system reaches a steady-state value different from the desired reference such that the constraint (4.10e) is active. It is shown that, under this situation, $\mathcal{X}_i^f$ is an invariant set according to Remark 4.6.

Remark 4.6 For the CMPC controller with limited available resource $\pi \in \mathbb{R}_{>0}$, suppose that a sub-system reaches a steady state at a time instant $k \in \mathbb{Z}_{\geq 0}$ denoted by $\mathbf{x}_{i,k}^s$. Notice that, if the steady state is achieved at the time instant $k \in \mathbb{Z}_{\geq 0}$, then $\mathbf{x}_{i,k+j} = \mathbf{x}_i^s$ using the control law $\mathbf{u}_{i,k+j} = \mathbf{u}_{i,k}^\star$, for all $j \in \mathbb{Z}_{\geq 0}$.   ◇

4.3.2 Topologies and Proposed Non-centralized MPC Control Design

The CMPC topology is composed by an MPC controller that collects the information from all the sub-systems and then sends the corresponding control inputs according to the solution of the optimization problem (4.10). In contrast, the topology for the DMPC controller with population dynamics implies a local MPC controller per sub-system. Each local controller needs to solve a partial problem and then coordinates its control signal by means of the DSD[3] considering the constraint (4.10e).

If the constraint (4.10e) is omitted, then the optimization problem (4.10) can be decoupled since dynamics of the sub-systems are decoupled as well as constraints (4.10b), (4.10c), and (4.10d). Consequently, a local MPC controller for the ith sub-system is designed by considering an optimization problem denoted by $\mathcal{P}_{\text{LMPC}}$ as follows:

$$\underset{\mathbf{u}_{i,k|k},\dots,\mathbf{u}_{i,k+H_p-1|k}}{\text{minimize}} \quad J_{i,k} = J_i^f\left(\mathbf{x}_{i,k+H_p|k}\right) + \sum_{j=0}^{H_p-1} J_i^\ell\left(\mathbf{x}_{i,k+j|k}, \mathbf{u}_{i,k+j|k}\right), \qquad (4.13)$$

subject to (4.10b), (4.10c), and (4.10d). At time instant k, these local MPC controllers compute the optimal sequences $\hat{\mathbf{u}}_{i,k}^\star$ for all sub-systems $i \in \tilde{\mathcal{S}}$, from which $\mathbf{u}_{i,k}^\star$ is obtained. In order to deal with the constraint (4.10e), a distributed full-potential game with DSD is proposed taking advantage of the optimization problem form in

---

[3]DSD have been selected to illustrate the methodology. However, any of the distributed population dynamics can be used.

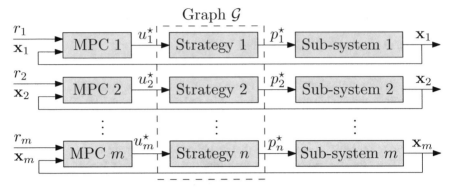

Fig. 4.6 Non-centralized MPC with distributed population dynamics scheme to solve the problem in (4.10) (taken from [2])

(2.13) and Theorem 2.1. For simplicity, and without loss of generality, it is considered from now on that each sub-system has only one control input, i.e., $n_{u,i} = 1$, for all $i \in \tilde{S}$, then the optimal control input computed by the local MPC controller of the sub-system $i \in \tilde{S}$ at time instant k is denoted by $u_{i,k}^\star$. The non-centralized scheme is shown in Fig. 4.6. Since (4.10e) is not an equality constraint, it is necessary to add a slack variable, denoted by p_{n+1}, to the game. This slack variable is treated as a new node added to the graph and can be connected to any other arbitrary node. Additionally, its fitness function is chosen as $f_{n+1} = 0$. The slack variable allows the controller to use less than the total available resource when it is convenient.

Notice that the solution of the LMPC with $n_{u,i} = 1$ for all $i \in \tilde{S}$, denoted as $u_{i,k}^\star$, is an input to the DSD, which computes in a distributed way the final optimal control input $p_i^\star$ applied to the associated sub-system. Furthermore, the local MPC controller supplies the bounds $u_{i,k}^{min}, u_{i,k}^{max} \in \mathbb{R}$ for the corresponding control signal, such that the problem (4.13) is feasible.

For an arbitrary number of control inputs, the bound $\mathbf{u}_{i,k}^{min} \in \mathbb{R}^{n_{u,i}}$ is found with local information by solving the optimization problem (4.13) with weights $\mathbf{Q}_i = \mathbf{0}_{n_{x,i} \times n_{x,i}}$ and $\mathbf{R}_i = \mathbb{I}_{n_{u,i}}$; and the bound $\mathbf{u}_{i,k}^{max} \in \mathbb{R}^{n_{u,i}}$ is similarly found with $\mathbf{Q}_i = \mathbb{I}_{n_{x,i}}$ and $\mathbf{R}_i = \mathbf{0}_{n_{u,i} \times n_{u,i}}$. Both problems are solved subject to (4.10b), (4.10c), and (4.10d).

Remark 4.7 Generally, the bounds $\underline{\mathbf{u}}_i$ and $\bar{\mathbf{u}}_i$ in (4.10d) are different from $\mathbf{u}_{i,k}^{min}$ and $\mathbf{u}_{i,k}^{max}$, respectively. The values $\underline{\mathbf{u}}_i$ and $\bar{\mathbf{u}}_i$ determine the physical constraints for the control inputs, whereas $\mathbf{u}_{i,k}^{min}$ and $\mathbf{u}_{i,k}^{max}$ determine the bounds of control inputs that guarantee feasibility of the optimization problem (4.13). $\diamond$

A strictly-concave full-potential function is proposed for the distributed population dynamics. This function corresponds to the objective function of Problem (2.13), which is denoted by $\mathcal{P}_{DSD}$. For simplicity in the notation, the discrete-time sub-index for the control input $u_i^\star$, and for bounds u_i^{min} and u_i^{max}, is omitted. The full-potential function is written as follows [1]:

$$V(\mathbf{p}) = -\sum_{i=1}^{n} w_i (u_i^\star - p_i)^2, \tag{4.14}$$

where w_i assigns a weighting factor to each control input, e.g., if $w_i = |x_i - r_i|/r_i$, for all $i \in \mathcal{S}$, then more priority is assigned to those sub-systems with higher percentage error. Notice that this selection for w_i requires that $r_i > 0$, for all $i \in \mathcal{S}$, and alternative potential functions $V(\mathbf{p})$ can be used satisfying that their maximum are obtained when $u_i^\star = p_i$. Consequently, the fitness functions for the game are given by $\mathbf{f}(\mathbf{p}) = \nabla V(\mathbf{p})$, i.e., $f_i(p_i) = 2w_i(u_i^\star - p_i)$. Note that this methodology does not require full information of all control inputs and/or all states of sub-systems since:

(i) the graph $\mathcal{G}$ representing the information interaction among sub-systems and all control inputs is a non-complete graph, and

(ii) the proposed fitness functions are decoupled, i.e., f_i depends only on information of the ith sub-system.

In order to satisfy the feasible region of the optimization problem (4.10), the DSD are modified as follows:

$$\dot{p}_i = \sum_{j \in \mathcal{N}_i} \left(p_j - u_j^{\min} \right) \left[f_i(\mathbf{p}) - f_j(\mathbf{p}) \right]_+ - \left(p_i - u_i^{\min} \right) \sum_{j \in \mathcal{N}_i} \left[f_j(\mathbf{p}) - f_i(\mathbf{p}) \right]_+, \ \forall i \in \mathcal{S}. \tag{4.15}$$

Remark 4.8 It is known that $u_i^{\min} \le u_i^\star \le u_i^{\max}$. There are two possible cases:

(i) when there is enough resource in the system, $p_i^\star = u_i^\star$ and the final control input belongs to the feasible region, and

(ii) when there is not enough resource in the system, then $p_i^\star < u_i^\star$ and the term $p_i - u_i^{\min}$ in (4.15) guarantees that the evolution of p_i will never be under the value of $u_i^{\min}$, due to the fact that $\sum_{j \in \mathcal{N}_i} \left(p_j - u_j^{\min} \right) \left[f_i(\mathbf{p}) - f_j(\mathbf{p}) \right]_+ \ge 0$ in (4.15).

In conclusion, the final control input $p_i^\star$ belongs to the feasible region of the optimization problem in (4.10). ◇

Each sub-system has a local MPC controller in which the optimization problem (4.13) is solved every $k \in \mathbb{Z}_{\ge 0}$, then there is a set of n controllers generating an optimal control input $u_{i,k}^\star$ for all $i \in \mathcal{S}$. This optimal control input (with respect to (4.13)) provides a fitness function $f_i(p_i)$ to the DSD that compute, in a distributed way, the final control input $p_i^\star$ satisfying the constraint (4.10e), and the same procedure can be applied for the whole optimal sequence of control inputs.

Remark 4.9 In order to initialize the DSD, the initial conditions at each iteration should belong to the feasible region supplied also by the local MPC controller. To establish this initial condition, an algorithm for solving the associated constraint satisfaction problem (CSP) should be implemented [23]. Moreover, notice that due to

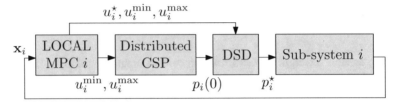

Fig. 4.7 Flow diagram of the proposed methodology with $n_{u,i} = 1$, for all sub-systems (taken from [2])

the fact that not all sub-systems can communicate with each other, the CSP algorithm should be performed in a distributed manner.   ◇

Figure 4.7 shows the flow diagram of the proposed population-games-based DMPC with DSD. The DSD require $u_i^\star$, for all $i \in \mathcal{S}$, to set the fitness functions in the game. The DSD also require the limits $[u_i^{\min}, u_i^{\max}]$, for all $i \in \mathcal{S}$, in order to guarantee that the vector of final control inputs $\mathbf{p}^\star$ belongs to the feasible regions, i.e., $u_i^{\min} \leq p_i^\star \leq u_i^{\max}$, for all $i \in \mathcal{S}$. Moreover, the CSP must be solved in a distributed way since there is not full information in the distributed configuration.

4.3.3 Plug-and-Play Property

One of the advantages of the distributed control design with population dynamics is the reduction of information dependence. Furthermore, there is another relevant advantage associated to the proposed scheme. The methodology consists in dividing the original problem into different sub-problems whose solutions are coordinated to obtain a final control input. In this regard, each control problem associated to each sub-system is independent from the others.

Now, suppose that a new sub-system is added to the initial problem (4.10), i.e., that the number of sub-systems is now $n + 1$, and then only sum upper limits involved in (4.10a) and (4.10e) should be modified in the MPC optimization problem. Notice that for this new optimization problem, the decoupled set of optimization problems is the same as in (4.13), including the optimization problem associated to the sub-system $n + 1$, and the bounds $u_{n+1}^{\min}$ and $u_{n+1}^{\max}$ can be found without requiring information from the other sub-systems. Finally, a new node is added to the graph $\mathcal{G}$ and the CSP computes a feasible initial condition for the DSD.

Consequently, the proposed control scheme shows the plug-and-play feature since it is not necessary to modify previously already designed parts of the MPC controller in order to add a new sub-system to the problem.

Remark 4.10 The same analysis may be performed for the removal of sub-systems to the problem, but it should be taken into account that the graph $\mathcal{G}$ cannot become disconnected after this modification.   ◇

4.3.4 Control Convergence Cases

Once the methodology has been presented, this subsection is dedicated to analyze two possible cases that might occur when computing the optimal control input in a distributed way with the proposed methodology. Notice that the adjacency matrix $\mathbf{A}^{(\mathbf{p})}$ of the graph $\mathcal{G}^{(\mathbf{p})}$ depends on the proportion of agents in the population game. Consequently, when a constraint within the population-dynamics optimization problem is active, i.e., $p_i = u_i^{\min}$ or $p_i = u_i^{\max}$, this might cause the disconnection of the graph (an element within the adjacency matrix $\mathbf{A}^{(\mathbf{p})}$ becomes null). However, this fact also depends on the topology of the graph $\mathcal{G}$.

Case 1: $u_i^{\star} > u_i^{\min}$, for all $i \in \mathcal{S}$. It is known that the optimal point $\mathbf{p}^{\star} \in \mathrm{int}\Delta$, the graph $\mathcal{G}^{(\mathbf{p})}$ is connected for all t since $\mathbf{u}^{\star}$ is an interior point of $\mathcal{U}$, and Theorem 4.2 holds.

Case 2: $u_i^{\star} = u_i^{\min}$, for some $i \in \mathcal{S}$. The optimal control input $\mathbf{u}^{\star}$ is at the edge of $\mathcal{U}$, i.e., there is an active constraint. Consequently, the node associated to that decision variable disappears and $\mathcal{G}^{(\mathbf{p})}$ might get disconnected depending on its topology. Then, each problem in each sub-complete graph $\mathcal{G}' \subset \mathcal{G}^{(\mathbf{p})}$ converges to an optimal solution, and the global problem is solved getting a sub-optimal solution. However, an appropriate design of redundant links might solve this issue.

4.3.5 Constraint Satisfaction Problem

As stated in Theorem 4.1, an important feature of the DSD given in (4.15) is that they guarantee constraint satisfaction along the time provided that the initial condition belongs to the feasible region of the considered problem. Therefore, a feasible initial condition should be found in order to initialize the DSD, i.e., $\mathbf{p}(0)$ must satisfy the following constraints:

$$u_i^{\min} \leq p_i(0) \leq u_i^{\max}, \quad \text{(feasibility of the local MPC controller)}, \tag{4.16a}$$

$$p_{n+1}(0) \geq 0, \quad \text{(positivity of the slack variable)}, \tag{4.16b}$$

$$\sum_{i=1}^{n+1} p_i(0) = \pi, \quad \text{(resource constraint)}. \tag{4.16c}$$

The above requirements are not trivial since a distributed framework in which each node of the network only has partial information of the whole problem is considered. This section describes a method (inspired by the algorithm proposed in [24], where the node counting has been presented, among several other applications that do not include the CSP) that solves the CSP characterized by the constraints (4.16) in a distributed way. The CSP is solved before applying the DSD.

First of all, consider the information available at each node: (*i*) the ith node knows its local bounds, i.e., $u_i^{\min}$ and $u_i^{\max}$; (*ii*) it is assumed that the slack node knows the available resource π; and (*iii*) nodes can share information by using the communication network that is given by the connected graph $\mathcal{G}$. Assuming that there exists a vector $\mathbf{p}(0)$ that satisfies the constraints in (4.16), a possible choice for $\mathbf{p}(0)$ is given as follows:

$$p_i(0) = u_i^{\min}, \quad \forall i \in \mathcal{S},$$
$$p_{n+1}(0) = \pi - \xi, \tag{4.17}$$

where $\xi = \sum_{i \in \mathcal{S}} u_i^{\min}$. Notice that this solution can be computed directly by using only local information except for ξ. Therefore, the idea is that the $(n+1)$th node obtains ξ by means of a distributed algorithm. In order to do so, ξ is rewritten as the product of the total number of nodes by the average of the minimum boundaries of nodes, i.e.,

$$\xi = \sum_{i \in \mathcal{S}} u_i^{\min},$$
$$= (n+1) \, \mathrm{mean}(u_1^{\min}, \ldots, u_n^{\min}, 0), \tag{4.18}$$

where $\mathrm{mean}(\cdot)$ denotes the arithmetic mean. Now, the original problem is divided into two sub-problems: (*i*) find the average of the lower bounds of nodes; and (*ii*) find the total number of nodes. Notice that each problem needs a distributed solution.

Finding the Average of the Lower Bounds of Nodes

The main idea is that the information of nodes about lower bounds propagates through the network. For this purpose, an auxiliary variable $\xi_i^{\min} \in \mathbb{R}$ per node is defined, where the sub-index i denotes that the variable is associated with the ith node. These variables are initialized with the corresponding node lower bound as follows: $\xi_i^{\min}(0) = u_i^{\min}$, $\forall i \in \mathcal{S}$, and $\xi_{n+1}^{\min}(0) = 0$. Notice that the arithmetic mean of the lower bounds of nodes is equal to the arithmetic mean of the initial conditions of the auxiliary variables defined above, i.e.,

$$\mathrm{mean}(u_1^{\min}, \ldots, u_n^{\min}, 0) = \mathrm{mean}(\xi_1^{\min}(0), \ldots, \xi_{n+1}^{\min}(0)).$$

In order to calculate this quantity in a distributed way, a standard average consensus algorithm as

$$\dot{\xi}_i^{\min} = \sum_{j \in \mathcal{N}_i} \left(\xi_j^{\min} - \xi_i^{\min} \right), \tag{4.19}$$

can be applied [6]. According to [6], if the communication network is described by a connected graph, then $\xi_i^{\min\star} = \mathrm{mean}(\xi_1^{\min}(0), \ldots, \xi_{n+1}^{\min}(0))$, for all $i = 1, \ldots, n+1$, where $\xi_i^{\min\star}$ is the steady-state value of $\xi_i^{\min}$. Thus, $\xi_i^{\min\star} = \mathrm{mean}(u_1^{\min}, \ldots, u_n^{\min}, 0)$.

This implies that the $(n + 1)$th node is capable to obtain the required value by using only local information.

Finding the Total Number of Nodes

The second problem is to locally compute the total number of nodes. In order to do so, a similar procedure as in the previous problem is followed. Define another auxiliary variable per node. Let ξ_i^c be the variable associated to the ith node. These new auxiliary variables are initialized as follows: $\xi_i^c(0) = 0$, for all $i \in \mathcal{S}$, and $\xi_{n+1}^c(0) = 1$. The above initialization values are important since their average is related to the required information, i.e., $\text{mean}(\xi_1^c(0), \ldots, \xi_{n+1}^c(0)) = (n + 1)^{-1}$. Thus, the same algorithm as in (4.19) can be applied to compute the needed quantity in a distributed way, i.e.,

$$\dot{\xi}_i^c = \sum_{j \in \mathcal{N}_i} \left(\xi_j^c - \xi_i^c \right). \tag{4.20}$$

Again, $\xi_i^{c\star} = \text{mean}(\xi_1^c(0), \ldots, \xi_{n+1}^c(0)) = (n + 1)^{-1}$, for all $i = 1, \ldots, n + 1$, in steady-state. Therefore, the $(n + 1)$th node can get the required quantity by taking $(\xi_{n+1}^{c\star})^{-1}$.

Summarizing, the steady-state solutions of (4.19)–(4.20) are used to compute ξ in (4.18), i.e., $\xi = \xi_i^{\min\star}/\xi_i^{c\star}$, and then this result is replaced in (4.17) in order to obtain the required initial feasible point for the DSD. Notice that this procedure is performed using only local information.

4.4 Stability Analysis of Population-Games-Based DMPC

After having presented the proposed population-games-based DMPC controller, which is composed of three different main stages, i.e., m local MPC controllers, a distributed CSP, and a full-potential game with DSD, it is necessary to develop the stability analysis for the proposed approach. The stability analysis of the population-games-based DMPC is presented in Theorem 4.4.

Theorem 4.4 *The system (4.9) is stabilized by the proposed population-games-based DMPC controller with invariant region of attraction $\mathcal{X}_i^A = \left\{ \mathbf{x}_i \in \mathcal{X}_i : J_i^c(\mathbf{x}_{i,k+H_p}^\star) \leq \pi_i^c \right\}$, for all the sub-systems $i = 1, \ldots, m$, and considering that*

1. *$\mathbf{x}_{i,0} \in \mathcal{X}_i$, for all the sub-systems $i = 1, \ldots, m$, and*
2. *Assumptions 4.3 and 4.4 hold for all the sub-systems $i = 1, \ldots, m$.*

Furthermore, given a centralized MPC with parameters $\mathbf{Q}_i$, $\mathbf{R}_i$, $\mathbf{V}_i$, and references $\mathbf{r}_i$, for all $i = 1, \ldots, m$, there is an equivalent population-games-based DMPC with parameters $\hat{\mathbf{Q}}_{i,k} = \psi_i \mathbf{Q}_i$, $\hat{\mathbf{R}}_{i,k} = \psi_i \mathbf{R}_i$, $\hat{\mathbf{V}}_{i,k} = \psi_i \mathbf{V}_i$, *where* $\psi_i > 0$ *and the weights for fitness functions at each sub-system* $\omega_i = \left[\mathbf{B}_{d,i}^\top \hat{\mathbf{Q}}_{i,k} \mathbf{B}_{d,i} + \hat{\mathbf{R}}_{i,k} \right] \in \mathbb{R}^{n_{u,i} \times n_{u,i}}$, *for all* $i = 1, \ldots, m$.

Proof (Scketch) This proof is divided into two parts. First, it is shown that there exists a CMPC whose solution is equivalent to the solution obtained with the proposed DMPC. Then, the equivalence allows to derive stability conditions of the closed-loop system controlled using the population-games-based DMPC.

First part: It is shown that there exists a relation between the prioritization of the MPC controller in the cost function (4.10a), i.e., values of $\mathbf{Q}_i$, $\mathbf{V}_i$, and $\mathbf{R}_i$, with $i = 1, \ldots, m$, and the prioritization of the DSD in the potential function, i.e., w_i, with $i = 1, \ldots, n$, where $n = \sum_{j=1}^m n_{u,j}$ (one weight w_i for each control input since $\mathbf{u}_i \in \mathbb{R}^{n_{u,i}}$) such that the optimal control input $\mathbf{u}^\star$ obtained from the solution of the centralized optimization problem $\mathcal{P}_{\text{MPC}}$ is the same as the optimal control input $\mathbf{p}^\star$ obtained from the multi-stage distributed strategy through the optimization problems $\mathcal{P}_{\text{LMPC}}$ and $\mathcal{P}_{\text{DSD}}$. For simplicity and without loss of generality, a constant reference for each sub-system is considered, i.e., $\mathbf{r}_{i,k} = \mathbf{r}_i$, for all k.

Consider the vectors $\mathbf{X}_{i,k} = \begin{bmatrix} \mathbf{x}_{i,k+1|k}^\top & \mathbf{x}_{i,k+2|k}^\top & \cdots & \mathbf{x}_{i,k+H_p|k}^\top \end{bmatrix}^\top \in \mathbb{R}^{n_{x,i} H_p}$, and $\mathbf{U}_{i,k} = \begin{bmatrix} \mathbf{u}_{i,k|k}^\top & \mathbf{u}_{i,k+1|k}^\top & \cdots & \mathbf{u}_{i,k+H_p-1|k}^\top \end{bmatrix}^\top \in \mathbb{R}^{n_{u,i} H_p}$. Then, the prediction model can be written as $\mathbf{X}_{i,k} = \boldsymbol{\Psi}_i \mathbf{x}_{i,k|k} + \boldsymbol{\Theta}_i \mathbf{U}_{i,k}$, and constraints of the form $\underline{\mathbf{x}}_i \leq \mathbf{x}_i \leq \bar{\mathbf{x}}_i$, and $\underline{\mathbf{u}}_i \leq \mathbf{u}_i \leq \bar{\mathbf{u}}_i$, may be compacted as $\boldsymbol{\Xi}_i \mathbf{U}_{i,k} \leq \boldsymbol{\xi}_{i,k}$. In order to determine the cost function in its matricial form, consider $\boldsymbol{\Phi}_i = [\mathbf{r}_i^\top \cdots \mathbf{r}_i^\top]^\top$, i.e.,

$$J_k = \sum_{i=1}^m J_{i,k},$$

$$J_{i,k} = \left(\mathbf{X}_{i,k} - \boldsymbol{\Phi}_i \right)^\top \tilde{\mathbf{Q}}_i \left(\mathbf{X}_{i,k} - \boldsymbol{\Phi}_i \right) + \mathbf{U}_{i,k}^\top \tilde{\mathbf{R}}_i \mathbf{U}_{i,k}.$$

Defining $\mathbf{E}_{i,k} = \boldsymbol{\Phi}_i - \boldsymbol{\Psi}_i \mathbf{x}_{i,k|k}$, $\mathbf{G}_{i,k} = 2\boldsymbol{\Theta}_i^\top \tilde{\mathbf{Q}}_i \mathbf{E}_{i,k}$, and $\mathbf{H}_i = \boldsymbol{\Theta}_i^\top \tilde{\mathbf{Q}}_i \boldsymbol{\Theta}_i + \tilde{\mathbf{R}}_i$, the cost function is written as[4]

$$J_{i,k} = \mathbf{U}_{i,k}^\top \mathbf{H}_i \mathbf{U}_{i,k} - \mathbf{U}_{i,k}^\top \mathbf{G}_{i,k} + \mathbf{E}_{i,k}^\top \tilde{\mathbf{Q}}_i \mathbf{E}_{i,k}.$$

Besides, consider the expression

$$\underbrace{\begin{bmatrix} \mathbb{1}_{n_{u,i}}^\top & & \mathbf{0} \\ & \ddots & \\ \mathbf{0} & & \mathbb{1}_{n_{u,i}}^\top \end{bmatrix}}_{\alpha_i} \mathbf{U}_{i,k} = \begin{bmatrix} \mathbb{1}_{n_{u,i}}^\top \mathbf{u}_{i,k|k} \\ \vdots \\ \mathbb{1}_{n_{u,i}}^\top \mathbf{u}_{i,k+H_p-1|k} \end{bmatrix}.$$

---

[4] Notice that $\mathbf{E}_{i,k}$ is a constant known value at each iteration since $\boldsymbol{\Phi}_i$ and $\boldsymbol{\Psi}_i$ are constant and the current system state $\mathbf{x}_{i,k|k}$ is also known for all $i = 1, \ldots, m$. Therefore, $\mathbf{G}_{i,k}$ is also constant at each iteration.

(1) The optimization problem $\mathcal{P}_{\mathrm{MPC}}$ behind the MPC controller is stated as follows:

$$\underset{\mathbf{U}_{i,k}}{\text{minimize}}\; J_k = \sum_{i=1}^{m} \left\{ \mathbf{U}_{i,k}^{\top} \mathbf{H}_i \mathbf{U}_{i,k} - \mathbf{U}_{i,k}^{\top} \mathbf{G}_{i,k} + \mathbf{E}_{i,k}^{\top} \tilde{\mathbf{Q}}_i \mathbf{E}_{i,k} \right\}, \qquad (4.21\text{a})$$

subject to

$$\Xi_i \mathbf{U}_{i,k} \le \boldsymbol{\xi}_{i,k}, \quad \forall\, i = 1, \ldots, m, \qquad (4.21\text{b})$$

$$\sum_{j=1}^{m} \alpha_j \mathbf{U}_{j,k} \le \mathbb{1}_{[H_p]} \pi. \qquad (4.21\text{c})$$

For the optimization problem in (4.21), the following Lagrangian function is defined:

$$\bar{J}_k = \sum_{i=1}^{m} \left\{ J_{i,k} + \boldsymbol{\mu}_{i,k}^{\top} \left[\Xi_i \mathbf{U}_{i,k} - \boldsymbol{\xi}_{i,k} \right] \right\} + \boldsymbol{\varepsilon}_k^{\top} \left[\sum_{j=1}^{m} \alpha_j \mathbf{U}_{j,k} - \mathbb{1}_{[H_p]} \pi \right],$$

where $\boldsymbol{\mu}_i$, and $\boldsymbol{\varepsilon}$ are associated to the Lagrange multipliers. Then, the corresponding KKT conditions for each sub-system are written as follows:

$$2\mathbf{H}_i \mathbf{U}_{i,k}^{\star} - \mathbf{G}_{i,k} + \boldsymbol{\mu}_{i,k}^{\top} \Xi_i + \boldsymbol{\varepsilon}_k^{\top} \alpha_i = \mathbf{0}, \qquad (4.22\text{a})$$

$$\Xi_i \mathbf{U}_{i,k}^{\star} \le \boldsymbol{\xi}_{i,k}, \qquad (4.22\text{b})$$

$$\sum_{j=1}^{m} \alpha_j \mathbf{U}_{j,k}^{\star} \le \mathbb{1}_{[H_p]} \pi, \qquad (4.22\text{c})$$

$$\boldsymbol{\mu}_{i,k}^{\top} \left[\Xi_i \mathbf{U}_{i,k}^{\star} - \boldsymbol{\xi}_{i,k} \right] = \mathbf{0}, \qquad (4.22\text{d})$$

$$\boldsymbol{\varepsilon}_k^{\top} \left[\sum_{j=1}^{m} \alpha_j \mathbf{U}_{j,k}^{\star} - \mathbb{1}_{[H_p]} \pi \right] = \mathbf{0}, \qquad (4.22\text{e})$$

$$\boldsymbol{\mu}_{i,k}, \boldsymbol{\varepsilon}_k \ge \mathbf{0}. \qquad (4.22\text{f})$$

(2) When the minimization of the costs functions for the sub-systems considers the coupled constraint, more importance is assigned to those sub-systems with higher errors and with more prioritization weights. However, when the coupled constraint is neglected, the prioritization in the multi-objective cost function is lost. In order to take into account this effect, an auxiliary weight ψ_i^{-1} with $\psi_i > 0$ is considered at each decoupled $J_{i,k}$ for maintaining the original prioritization throughout the proof. Therefore, the cost function corresponding to the local MPC controller is given by

$$J_{i,k} = \left(\mathbf{X}_{i,k} - \boldsymbol{\Phi}_i \right)^{\top} \psi_i^{-1} \tilde{\mathbf{Q}}_i \left(\mathbf{X}_{i,k} - \boldsymbol{\Phi}_i \right) + \mathbf{U}_{i,k}^{\top} \psi_i^{-1} \tilde{\mathbf{R}}_i \mathbf{U}_{i,k}.$$

Notice that the addition of ψ_i^{-1} does not modify the optimal point of $J_{i,k}$. Then, the optimization problem behind the local MPC controllers $\mathcal{P}_{\text{LMPC}}$ are stated as follows:

$$\underset{\hat{\mathbf{U}}_{i,k}}{\text{minimize}} \ J_{i,k} = \hat{\mathbf{U}}_{i,k}^{\top} \hat{\mathbf{H}}_i \hat{\mathbf{U}}_{i,k} - \hat{\mathbf{U}}_{i,k}^{\top} \hat{\mathbf{G}}_{i,k} + \mathbf{E}_{i,k}^{\top} \tilde{\mathbf{Q}}_i \mathbf{E}_{i,k}, \qquad (4.23a)$$

subject to

$$\mathbf{\Xi}_i \hat{\mathbf{U}}_{i,k} \le \boldsymbol{\xi}_{i,k}, \qquad (4.23b)$$

where $\hat{\mathbf{G}}_{i,k} = 2\mathbf{\Theta}_i^{\top} \psi_i^{-1} \tilde{\mathbf{Q}}_i \mathbf{E}_{i,k}$, and $\hat{\mathbf{H}}_i = \mathbf{\Theta}_i^{\top} \psi_i^{-1} \tilde{\mathbf{Q}}_i \mathbf{\Theta}_i + \psi_i^{-1} \tilde{\mathbf{R}}_i$ (matrices $\tilde{\mathbf{Q}}_i$, and $\tilde{\mathbf{R}}_i$ are selected in a way that $\mathbf{\Theta}_i^{\top} \psi_i^{-1} \tilde{\mathbf{Q}}_i \mathbf{\Theta}_i + \psi_i^{-1} \tilde{\mathbf{R}}_i > 0$ to ensure that $\hat{\mathbf{H}}_i^{-1}$ exists [25]). In order to avoid confusion between the control inputs from the centralized MPC controller and those from the local MPC controllers, the optimal output of the local MPC controller is denoted by $\hat{\mathbf{U}}_i^{\star}$. The Lagrangian function is written as follows:

$$\bar{J}_{i,k} = J_{i,k} + \boldsymbol{\lambda}_{i,k}^{\top} \left[\mathbf{\Xi}_i \hat{\mathbf{U}}_{i,k} - \boldsymbol{\xi}_k \right],$$

where $\boldsymbol{\lambda}_i$ are associated to the Lagrange multipliers. Then, the corresponding KKT conditions are written as follows:

$$2\hat{\mathbf{H}}_i \hat{\mathbf{U}}_{i,k}^{\star} - \hat{\mathbf{G}}_{i,k} + \boldsymbol{\lambda}_{i,k}^{\top} \mathbf{\Xi}_i = \mathbf{0}, \qquad (4.24a)$$

$$\mathbf{\Xi}_i \hat{\mathbf{U}}_{i,k}^{\star} \le \boldsymbol{\xi}_{i,k}, \qquad (4.24b)$$

$$\boldsymbol{\lambda}_{i,k}^{\top} \left[\mathbf{\Xi}_i \hat{\mathbf{U}}_{i,k}^{\star} - \boldsymbol{\xi}_{i,k} \right] = \mathbf{0}, \qquad (4.24c)$$

$$\boldsymbol{\lambda}_i(k) \ge \mathbf{0}. \qquad (4.24d)$$

(3) The variables of the optimization problem in the population dynamics are denoted by $\mathbf{q}_k$. The related problem is solved by using the DSD, i.e., $\mathbf{q}_k^{\star}$ is the Nash equilibrium of the full-potential population game. Consider $\mathbf{P}_{i,k} = [\mathbf{q}_{i,k|k}^{\top} \ \mathbf{q}_{i,k+1|k}^{\top} \ \cdots \ \mathbf{q}_{i,k+H_p-1|k}^{\top}]^{\top}$. Moreover, due to the fact that the proportion of agents is constrained by $\mathbf{u}_{i,k}^{\min} \le \mathbf{q}_{i,k} \le \mathbf{u}_{i,k}^{\max}$, then the optimization problem given by the population dynamics $\mathcal{P}_{\text{DSD}}$ is stated as follows:

$$\underset{\mathbf{P}_{i,k}}{\text{minimize}} \ f_k = \sum_{i=1}^{m} \left(\mathbf{P}_{i,k} - \hat{\mathbf{U}}_{i,k}^{\star} \right)^{\top} \mathbf{W}_i \left(\mathbf{P}_{i,k} - \hat{\mathbf{U}}_{i,k}^{\star} \right), \qquad (4.25a)$$

subject to

$$\Xi_i \mathbf{P}_{i,k} \leq \xi_{i,k}, \quad \text{for all } i = 1, \ldots, m, \tag{4.25b}$$

$$\sum_{j=1}^{m} \alpha_j \mathbf{P}_{j,k} \leq \mathbb{1}_{[H_p]} \pi, \tag{4.25c}$$

where the matrix $\mathbf{W}_i$, for all $i = 1, \ldots, m$, is non-singular. For the optimization problem in (4.25), the following Lagrangian function is defined:

$$\bar{f}_k = \sum_{i=1}^{m} \left\{ \left(\mathbf{P}_{i,k} - \hat{\mathbf{U}}_{i,k}^\star \right)^\top \mathbf{W}_i \left(\mathbf{P}_{i,k} - \hat{\mathbf{U}}_{i,k}^\star \right) + \right.$$

$$\left. \theta_{i,k}^\top \left[\Xi_i \mathbf{P}_{i,k} - \xi_k \right] \right\} + \beta_k^\top \left[\sum_{j=1}^{m} \alpha_j \mathbf{P}_{j,k} - \mathbb{1}_{[H_p]} \pi \right],$$

where θ_i, and β are associated to the Lagrange multipliers. Then, the corresponding KKT conditions, for all sub-systems $i = 1, \ldots, m$, are written as follows:

$$2\mathbf{W}_i \mathbf{P}_i^\star - 2\mathbf{W}_i \hat{\mathbf{U}}_{i,k}^\star + \theta_{i,k}^\top \Xi_i + \beta_k^\top \alpha_i = 0, \tag{4.26a}$$

$$\Xi_i \mathbf{P}_{i,k}^\star \leq \xi_{i,k}, \tag{4.26b}$$

$$\sum_{j=1}^{m} \alpha_j \mathbf{P}_{j,k}^\star \leq \mathbb{1}_{[H_p]} \pi, \tag{4.26c}$$

$$\theta_{i,k}^\top \left[\Xi_i \mathbf{P}_{i,k}^\star - \xi_{i,k} \right] = 0, \tag{4.26d}$$

$$\beta_k^\top \left[\sum_{j=1}^{m} \alpha_j \mathbf{P}_{j,k}^\star - \mathbb{1}_{[H_p]} \pi \right] = 0, \tag{4.26e}$$

$$\theta_{i,k}, \beta_k \geq 0. \tag{4.26f}$$

From (4.26a), it is obtained that

$$\hat{\mathbf{U}}_{i,k}^\star = \mathbf{P}_i^\star + \frac{1}{2} \mathbf{W}_i^{-1} \theta_{i,k}^\top \Xi_i + \frac{1}{2} \mathbf{W}_i^{-1} \beta_k^\top \alpha_i. \tag{4.27}$$

Now, replacing (4.27) in (4.24a), it is obtained

$$2\hat{\mathbf{H}}_i \mathbf{P}_i^\star - \hat{\mathbf{G}}_{i,k} + \left(\hat{\mathbf{H}}_i \mathbf{W}_i^{-1} \theta_{i,k}^\top + \lambda_{i,k}^\top \right) \Xi_i + \hat{\mathbf{H}}_i \mathbf{W}_i^{-1} \beta_k^\top \alpha_i = 0.$$

Notice that if the weights for the local MPC controllers (denoted by $\hat{\mathbf{Q}}_i$, $\hat{\mathbf{R}}_i$, and $\hat{\mathbf{V}}_i$) in the optimization problem are selected to be: $\hat{\mathbf{Q}}_i = \psi_i \mathbf{Q}_i$, $\hat{\mathbf{R}}_{i,k} = \psi_i \mathbf{R}_i$, and $\hat{\mathbf{V}}_{i,k} = \psi_i \mathbf{V}_i$, for all $i = 1, \ldots, m$, then $\hat{\mathbf{H}}_i = \mathbf{H}_i$, and $\hat{\mathbf{G}}_{i,k} = \mathbf{G}_{i,k}$. In addition, if $\mathbf{W}_i$ is selected to be $\hat{\mathbf{H}}_i$, for all $i = 1, \ldots, m$, then it follows that

$$2\mathbf{H}_i\mathbf{P}_i^\star - \mathbf{G}_{i,k} + \left(\boldsymbol{\theta}_{i,k} + \boldsymbol{\lambda}_{i,k}\right)^\top \boldsymbol{\Xi}_i + \boldsymbol{\beta}_k^\top \boldsymbol{\alpha}_i = \mathbf{0},$$

this condition is the same as (4.22a), and $\mathbf{P}_i^\star$ satisfies constraints (4.26b), and (4.26c). Therefore, the equivalence with the solution of the centralized optimization problem $\mathcal{P}_{\mathrm{MPC}}$ is shown. Due to the fact that only the first control input may be applied to the system, then it yields $\omega_i = \mathbf{B}_{d,i}^\top \hat{\mathbf{Q}}_{i,k} \mathbf{B}_{d,i} + \hat{\mathbf{R}}_{i,k}$, for all $i = 1, \ldots, m$.

Second part: Now, it is considered the optimal cost function of the CMPC controller as a Lyapunov function and it is proceed as in [19]. The cost is denoted by $J_k = \sum_{i=1}^m J_{i,k}$, and the optimal cost is denoted by $J_k^\star = \sum_{i=1}^m J_{i,k}^\star$. At time instant k, $\hat{\mathbf{u}}_{i,k}^\star = \left(\mathbf{u}_{i,k|k}^\star, \ldots, \mathbf{u}_{i,k+H_p-1|k}^\star\right)$ is the optimal control sequence for the ith subsystem. Similarly, the optimal control sequence at time instant $k+1$ for the ith sub-system is given by $\hat{\mathbf{u}}_{i,k+1}^\star = \left(\mathbf{u}_{i,k+1|k+1}^\star, \ldots, \mathbf{u}_{i,k+H_p|k+1}^\star\right)$. Furthermore, there are feasible control sequences given by

$$\hat{\mathbf{u}}_{i,k} = \left(\mathbf{u}_{i,k|k}^\star, \mathbf{u}_{i,k+1|k+1}^\star, \ldots, \mathbf{u}_{i,k+H_p-1|k+1}^\star\right), \text{ and,}$$

$$\hat{\mathbf{u}}_{i,k+1} = \left(\mathbf{u}_{i,k+1|k}^\star, \mathbf{u}_{i,k+2|k}^\star, \ldots, \mathbf{u}_{i,k+H_p-1|k}^\star, \mathbf{u}_{i\mathrm{MPC},k+H_p|k}\right).$$

The four previously introduced control sequences generate the costs $J_k^\star$, $J_{k+1}^\star$, J_k, and J_{k+1}, respectively. Then,

$$J_{i,k}^\star = J_i^f\left(\mathbf{x}_{i,k+H_p|k}\right) + \sum_{j=0}^{H_p-1} J_i^\ell\left(\mathbf{x}_{i,k+j|k}, \mathbf{u}_{i,k+j|k}^\star\right),$$

$$J_{i,k+1}^\star = J_i^f\left(\mathbf{x}_{i,k+H_p+1|k+1}\right) + \sum_{j=1}^{H_p} J_i^\ell\left(\mathbf{x}_{i,k+j|k+1}, \mathbf{u}_{i,k+j|k+1}^\star\right),$$

$$J_{i,k} = J_i^f\left(\mathbf{x}_{i,k+H_p|k+1}\right) + J_i^\ell\left(\mathbf{x}_{i,k|k}, \mathbf{u}_{i,k|k}^\star\right) + \sum_{j=1}^{H_p-1} J_i^\ell\left(\mathbf{x}_{i,k+j|k+1}, \mathbf{u}_{i,k+j|k+1}^\star\right),$$

$$J_{i,k+1} = J_i^f\left(\mathbf{x}_{i,k+H_p+1|k}\right) + \sum_{j=1}^{H_p-1} J_i^\ell\left(\mathbf{x}_{i,k+j|k}, \mathbf{u}_{i,k+j|k}^\star\right) + J_i^\ell\left(\mathbf{x}_{i,k+H_p|k}, \mathbf{u}_{i\mathrm{MPC},k+H_p|k}^\star\right).$$

Notice that $J_k^\star \leq J_k$, and $J_{k+1}^\star \leq J_{k+1}$. Consequently $J_{k+1}^\star + J_k^\star \leq J_{k+1} + J_k$, and $J_{k+1}^\star - J_k \leq J_{k+1} - J_k^\star$. The terms in the aforementioned inequality are

$$J_{k+1}^\star - J_k = \sum_{i=1}^m \left\{ J_i^\ell\left(\mathbf{x}_{i,k+H_p|k+1}, \mathbf{u}_{i,k+H_p|k+1}^\star\right) - J_i^\ell\left(\mathbf{x}_{i,k|k}, \mathbf{u}_{i,k|k}^\star\right) + \right.$$

$$\left. J_i^f\left(\mathbf{x}_{i,k+H_p+1|k+1}\right) - J_i^f\left(\mathbf{x}_{i,k+H_p|k+1}\right) \right\},$$

$$J_{k+1} - J_k^\star = \sum_{i=1}^m \left\{ J_i^\ell\left(\mathbf{x}_{i,k+H_p|k}, \mathbf{u}_{i\mathrm{MPC},k+H_p|k}^\star\right) - J_i^\ell\left(\mathbf{x}_{i,k|k}, \mathbf{u}_{i,k|k}^\star\right) \right.$$

$$\left. + J_i^f\left(\mathbf{x}_{i,k+H_p+1|k}\right) - J_i^f\left(\mathbf{x}_{i,k+H_p|k}\right) \right\}.$$

Replacing in the inequality and removing the term $J_i^\ell \left(\mathbf{x}_{i,k|k}, \mathbf{u}_{i,k|k}^\star \right)$ at both sides, it follows that:

$$\sum_{i=1}^m \left\{ J_i^\ell \left(\mathbf{x}_{i,k+H_p|k+1}, \mathbf{u}_{i,k+H_p|k+1}^\star \right) + J_i^f \left(\mathbf{x}_{i,k+H_p+1|k+1} \right) - J_i^f \left(\mathbf{x}_{i,k+H_p|k+1} \right) \right\}$$

$$\leq \sum_{i=1}^m \left\{ J_i^\ell \left(\mathbf{x}_{i,k+H_p|k}, \mathbf{u}_{i\mathbf{MPC},k+H_p|k}^\star \right) + J_i^f \left(\mathbf{x}_{i,k+H_p+1|k} \right) - J_i^f \left(\mathbf{x}_{i,k+H_p|k} \right) \right\}$$

$$\leq 0, \quad \text{(according to Assumption 4.3).} \tag{4.28}$$

The following steps of the proof are developed as the analysis presented in [19], i.e., suppose first that $\mathbf{x}_{i,k+H_p|k+1} = \mathbf{x}_i^s$, then the optimality implies that $\mathbf{x}_{i,k+H_p+1|k+1} = \mathbf{x}_i^s$ and it is concluded that $\mathbf{x}_{i,k+H_p+1|k+1} \in \mathcal{X}_i^f$. The other option occurs when $\mathbf{x}_{i,k+H_p|k+1} \neq \mathbf{x}_i^s$, in which case, using (4.28) and the fact $J_i^\ell \left(\mathbf{x}_{i,k+H_p|k}, \mathbf{u}_{i,k+H_p|k}^\star \right) > 0$, then

$$J_i^f \left(\mathbf{x}_{i,k+H_p+1|k+1} \right) < J_i^f \left(\mathbf{x}_{i,k+H_p|k+1} \right), \quad \forall \ i = 1, \ldots, m.$$

The only possibility to satisfy the strict inequality, and according to (4.11), is that $J_i^f \left(\mathbf{x}_{i,k+H_p+1|k+1} \right) = J_i^c < \pi_i^c$, and $J_i^f \left(\mathbf{x}_{i,k+H_p|k+1} \right) = \pi_i^c$, for all $i = 1, \ldots, m$. This fact implies that, according to the definition of the set $\mathcal{X}_i^A = \left\{ \mathbf{x}_{i,k} \in \mathcal{X}_i : J_i^c (\mathbf{x}_{i,k+H_p|k}) \leq \pi_i^c \right\}$, then $\mathbf{x}_{i,k+1|k+1} \in \mathcal{X}_i^A$. Hence, it is concluded that $\mathcal{X}_i^A$ is an invariant set. In order to prove the stability, it is recalled the fact that $J_{k+1}^\star \leq J_{k+1}$, which implies that $J_{k+1}^\star - J_k^\star \leq J_{k+1} - J_k^\star$, i.e.,

$$J_{i,k+1}^\star - J_{i,k}^\star \leq$$
$$J_i^f \left(\mathbf{x}_{i,k+H_p+1|k} \right) - J_i^f \left(\mathbf{x}_{i,k+H_p|k} \right) + J_i^\ell \left(\mathbf{x}_{i,k+H_p|k}, \mathbf{u}_{i\mathbf{MPC},k+H_p|k}^\star \right) - J_i^\ell \left(\mathbf{x}_{i,k|k}, \mathbf{u}_{i,k|k}^\star \right),$$

for all $i = 1, \ldots, m$. Using Assumption 4.3, it is concluded that

$$J_{k+1}^\star - J_k^\star \leq - \sum_{i=1}^m J_i^\ell \left(\mathbf{x}_{i,k|k}, \mathbf{u}_{i,k|k}^\star \right).$$

The cost function $J_k^\star$ is a decaying sequence. Hence the system (4.9) controlled by using a CMPC controller is stable. As a conclusion, according to the first part of the proof, there exists a CMPC controller whose solution is equal to the solution computed by the proposed population-games-based DMPC controller. Moreover, since the system controlled with the equivalent CMPC controller is stable according to the second part of the proof, then the system controlled with population-games-based DMPC is also stable. □

Corollary 4.1 *Given a CMPC with parameters* $\mathbf{Q}_i = \mathbf{Q}_j$, $\mathbf{V}_i = \mathbf{V}_j$, *and* $\mathbf{R}_i = \mathbf{R}_j$, *for all* $i, j = 1, \ldots, m$, *the equivalent population-games-based DMPC is obtained with the parameters* $\hat{\mathbf{Q}}_{i,k} = \psi_{i,k}\mathbf{Q}_i$, $\hat{\mathbf{R}}_{i,k} = \psi_{i,k}\mathbf{R}_i$, $\hat{\mathbf{V}}_i(k) = \psi_{i,k}\mathbf{V}_i$, *and* $\omega_i = \mathbf{B}_{d,i}^\top\hat{\mathbf{Q}}_{i,k}\mathbf{B}_{d,i} + \hat{\mathbf{R}}_{i,k}$, *with* $\psi_{i,k} = (\mathbf{r}_i - \mathbf{x}_{i,k|k})^\top(\mathbf{r}_i - \mathbf{x}_{i,k|k})$, *for all* $i = 1, \ldots, m$. $\diamond$

4.5 Case Study: Continuous Stirred Tank Reactor

The considered case study is an industrial process that comprises $m = 10$ continuously stirred tank reactors (CSTR) with constraints over the information interactions for the DMPC as presented in Fig. 4.8a. The vector of states for each sub-system is $\mathbf{x}_i = [C_i \ \ T_i]^\top$, where C_i is the concentration, and T_i is the temperature inside the CSTR, and the control input is given by $u_i = q_i$, where q_i corresponds to the inflow. This is a proof-of-concept problem to illustrate the performance of the proposed control approach. Moreover, the methodology is scalable to any higher dimension, i.e., for systems composed of any number of sub-systems. The control objective here is to maintain the concentrations $[1 \ \ 0]\mathbf{x}_1, [1 \ \ 0]\mathbf{x}_2, [1 \ \ 0]\mathbf{x}_3, \ldots, [1 \ \ 0]\mathbf{x}_9, [1 \ \ 0]\mathbf{x}_{10}$ as close as possible to the references $r_1 = 0.15, r_2 = 0.16, r_3 = 0.17, \ldots, r_9 = 0.23, r_{10} = 0.24$ mol/l, respectively. Additionally, the system has a limited total inflow resource for the control inputs $q_1, q_2, q_3, \ldots, q_{10}$ given by $\pi = 1500$ l/min. The physical constraints for the inflows are given by the range $[\underline{u}_i, \bar{u}_i] = [0, 300]$ (in l/min). It is assumed that there are local controllers guaranteeing that inflows achieve the values determined by the MPC controller, i.e., the proposed MPC controllers presented in this chapter compute the references to local controllers. The discrete-time linear models with a sampling time $\tau = 0.1$ min, for the first CSTR around the operational point given by $C_1' = 0.0823$ mol/l, $T_1' = 442$ K, and $q_1' = 100$ mol/l is written as follows:

$$\mathbf{x}_{1,k+1} = \begin{bmatrix} 0.33 & 1.29 \times 10^{-5} \\ 0.61 & 2.45 \times 10^{-5} \end{bmatrix} \mathbf{x}_{1,k} + \begin{bmatrix} 5.49 \times 10^{-4} \\ 1.95 \times 10^{-4} \end{bmatrix} u_{1,k},$$

$$\mathbf{y}_{1,k} = \begin{bmatrix} 1 & 0 \\ 0 & 1 \end{bmatrix} \mathbf{x}_{1,k}.$$

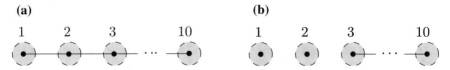

(a)   **(b)**

1   2   3   10   1   2   3   10

Fig. 4.8 Communication topology for the CSTRs case study. **a** The smaller possible connected graph representing a critic situation, and **b** The case when unplugging sub-systems 1 and 2

All the sub-systems have different model given by $\mathbf{A}_{d,j} = \mathbf{A}_{d,1}$
$1.01j$, $\mathbf{B}_{d,j} = \mathbf{B}_{d,1}1.01j$, for all $j = 2, \ldots, 10$. On the other hand, the weights
for the CMPC controller are given by $\mathbf{Q}_i = [10000 \ \ 0; 0 \ \ 0]$, and $\mathbf{R}_i = 1$, for all
$i = 1, \ldots, 10$.

Five different scenarios are presented, all of them with prediction horizon $H_p = 5$,
i.e.,

- **Scenario 1**: CMPC with weights $\mathbf{Q}_i$ and $\mathbf{R}_i$, and without resource constraint, i.e.,
  $\sum_{i=1}^{10} u_i \leq +\infty$.
- **Scenario 2**: CMPC with weights $\mathbf{Q}_i$ and $\mathbf{R}_i$, and with resource constraint, i.e.,
  $\sum_{i=1}^{10} u_i \leq 1800$.
- **Scenario 3**: Population-games-based DMPC with weights corresponding to The-
  orem 4.4, i.e., $\hat{\mathbf{Q}}_{i,k} = \psi_i \mathbf{Q}_i$, $\hat{\mathbf{R}}_{i,k} = \psi_i \mathbf{R}_i$, for all $i = 1, \ldots, 10$, with the weights
  in the population dynamics as in Theorem 4.4, i.e., $w_i = \mathbf{B}_{d,i}^\top \hat{\mathbf{Q}}_{i,k} \mathbf{B}_{d,i} + \hat{\mathbf{R}}_{i,k}$, with
  $\psi_{i,k} = (r_i - [1 \ \ 0]\mathbf{x}_{i,k|k})^2$, for all $i = 1, \ldots, 10$, and with resource constraint, i.e.,
  $\sum_{i=1}^{10} u_i \leq 1800$.
- **Scenario 4**: Population-games-based DMPC with local MPC weights $\hat{\mathbf{Q}}_i$ and $\hat{\mathbf{R}}_i$
  as in Scenario 3 with $\psi_i = 100(r_i - [1 \ \ 0]\mathbf{x}_{i,k|k})/r_i$, and with weights in the pop-
  ulation dynamics as $w_i = \psi_i$, for all $i = 1, \ldots, 10$, and with resource constraint,
  i.e., $\sum_{i=1}^{10} u_i \leq 1800$.
- **Scenario 5**: Population-games-based DMPC with local MPC with parameters as
  in Scenario 4, and with resource constraint, i.e., $\sum_{i=1}^{10} u_i \leq 1800$. Moreover, at the
  interval of time between 8 and 12 min, the sub-systems 1 and 2 are disconnected
  to illustrate the plug-and-play features of the proposed method (see Fig. 4.8b).

Results and Discussion

Figure 4.9 shows the evolution of system states (concentrations) for the ten CSTRs,
for all $i \in \tilde{S}$, and the five scenarios. As expected, the concentration of each CSTR
reaches its corresponding set-point when the total inflow is unconstrained. If this is not
the case (i.e., when the sum of inflows is limited to a value lower than 1800 l/min),
concentrations are below their corresponding set-points since there is not enough
feed-flow rate in the reactor mass balance to increase the concentration up to the
desired value. However, in the latter situation, controllers try to use all the available
resource to keep the controlled variables close to the desired value. Furthermore, for
the fifth scenario, two sub-systems are unplugged from the system and the remaining
seven sub-systems have more available resource achieving the required reference.
However, once the two sub-systems are plugged into the system, then the available
limited resource is newly optimally distributed. Figure 4.10 shows the evolution of
the control inputs for the different scenarios. Additionally, the equivalence between
the results obtained with Scenarios 2 and 3 as stated in Theorem 4.4 can be seen. It
can be seen that once the two sub-systems are unplugged from the system, then the

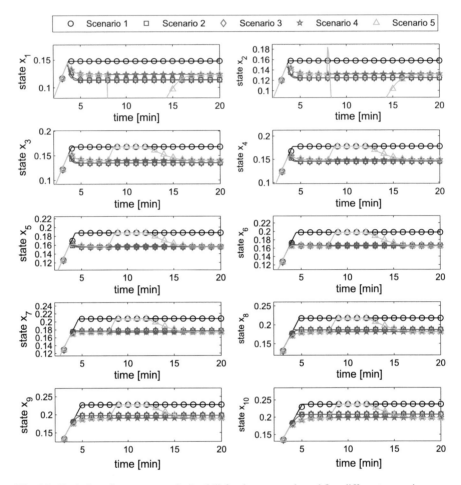

Fig. 4.9 Evolution of system states in [mol/l] for the case study and five different scenarios

resource that they were using is distributed throughout the other seven sub-systems, and afterwards, when the two sub-systems are plugged into the system, then the other seven sub-systems share the resource back again. Figure 4.11 shows the evolution of the total resource, where the satisfaction of the coupled constraint for all the scenarios is achieved.

Table 4.1 shows the steady-state error for the different scenarios considering the coupled constraint. The equivalence between the CMPC and the population-games-based DMPC can be seen. Besides, when weights w_i, for all $i \in \tilde{S}$, are selected to be the steady-state error for the population-games-based DMPC, then an evenhanded distribution of the resource is performed achieving a stationary state where all the percentages of error are the same. Consequently, the general distributed scheme can be tuned to the population game without modifying the local MPC controllers.

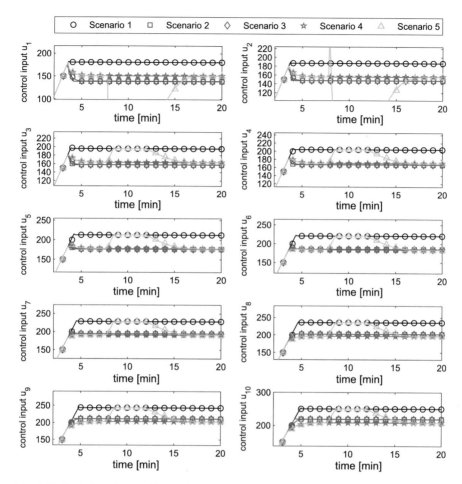

Fig. 4.10 Evolution of control inputs in [l/min] for the case study and five different scenarios

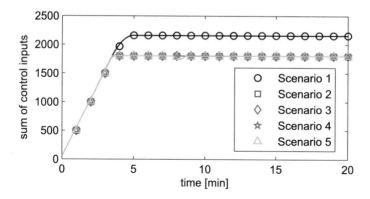

Fig. 4.11 Sum of control inputs for the five scenarios

Table 4.1 Steady-state error for the four scenarios

CSTR	Scenario 2 (%)	Scenario 3 (%)	Scenario 4 (%)
1	24.3	24.2	17.6
2	22.1	22.0	17.6
3	20.7	20.5	17.6
4	19.2	19.2	17.6
5	18.1	18.0	17.6
6	17.0	17.1	17.7
7	16.0	16.0	17.5
8	15.0	15.2	17.7
9	14.2	14.3	17.6
10	13.5	13.7	17.7

4.6 Summary

This chapter has proposed a novel methodology based on distributed population dynamics to make distributed an MPC scheme with a single coupled constraint associated to a limited resource. Results have shown that the methodology satisfies the coupled constraint in a distributed way. Also, throughout weighting parameters in the potential function, a dynamical tunning can be performed.

Simulations have shown that choosing the error as a weighting parameter, the same error is obtained in steady-state for all sub-systems. In order to make the problem non-centralized, a population games approach has been used. The DSD has been selected to design the DMPC controller. However, the same technique can be extended to other distributed population dynamics (e.g., distributed projection dynamics, distributed replicator dynamics, distributed logit choice dynamics, etc). Furthermore, since this distributed methodology is composed by different local and independent MPC controllers, the addition or removal of sub-systems to/from the entire system does not affect the local MPC controllers. The new design is obtained by modifying only the number of strategies in the DSD and guaranteeing communication of the new sub-system with at least another sub-system within the system.

Finally, it is pointed out that features of the distributed population games can be used for different control purposes besides the one presented in this chapter. To illustrate this, the next chapter presents a different control application in which the distributed population games can be used.

References

1. Barreiro-Gomez J, Obando G, Quijano N (2017) Distributed population dynamics: optimization and control applications. IEEE Trans Syst Man Cybern: Syst 47(2):304–314
2. Barreiro-Gomez J, Obando G, Ocampo-Martinez C, Quijano N (2015) Making non-centralized a model predictive control scheme by using distributed smith dynamics. In: Proceedings of the 5th IFAC conference on nonlinear model predictive control. Seville, Spain, pp 501–506
3. Sandholm WH (2010) Population games and evolutionary dynamics Mass. MIT Press, Cambridge
4. Pantoja A, Quijano N (2012) Distributed optimization using population dynamics with a local replicator equation. In: Proceedings of the 51st IEEE conference on decision and control (CDC). Maui, Hawaii, pp 3790–3795
5. Moreau L (2005) Stability of multiagent systems with time-dependent communication links. IEEE Trans Autom Control 50(2):169–182
6. Olfati-Saber R, Fax JA, Murray RM (2007) Consensus and cooperation in networked multi-agent systems. Proc IEEE 95(1):215–233
7. Ren W, Beard RW (2005) Consensus seeking in multiagent systems under dynamically changing interaction topologies. IEEE Trans Autom Control 50(5):655–661
8. Bornholdt S, Schuster HG (2003) Handbook of graphs and networks, 2nd edn. Wiley Online Library,
9. Maestre JM, Negenborn RR (2013) Distributed model predictive control made easy, vol 69. Springer Science and Business Media, Berlin
10. Boyd S, Parikh N, Chu E, Peleato B, Eckstein J (2011) Distributed optimization and statistical learning via the alternating direction method of multipliers. Found Trends Mach Learn 3(1):1–122
11. Mota J, Xavier J, Aguiar P, Puschel M (2013) D-ADMM: a communication-efficient distributed algorithm for separable optimization. IEEE Trans Signal Process 61(10):2718–2723
12. Costa RP, Lemos JM, Mota JFC, Xavier JMF (2014) D-ADMM based distributed MPC with input–output models. In: Proceedings of the 53rd IEEE conference on control applications (CCA). Antibes-Juan Les Pins, France, pp 699–704
13. Stewart BT, Venkat AN, Rawlings JB, Wright SJ, Pannocchia G (2010) Cooperative distributed model predictive control. Syst Control Lett 59(8):460–469
14. Ferramosca A, Limon D, Alvarado I, Camacho EF (2013) Cooperative distributed MPC for tracking. Automatica 49(2013):906–914
15. Trodden P, Richards A (2013) Cooperative distributed MPC of linear systems with coupled constraints. Automatica 49(2):479–487
16. Riverso S, Farina M, Ferrari-Trecate G (2013) Plug-and-play decentralized model predictive control for linear systems. IEEE Trans Autom Control 58(10):2608–2614
17. Riverso S, Farina M, Ferrari-Trecate G (2014) Plug-and-play model predictive control based on robust control invariant sets. Automatica 50(2014):2179–2186
18. Zeilinger MN, Y-Pu, Riverso S, Ferrari-Trecate G, Jones CN (2013) Plug and play distributed model predictive control based on distributed invariance and optimization. In: Proceedings of the 52nd IEEE conference on decision and control (CDC). Florenze, Italy, pp 5770–5776
19. Hu B, Linnemann A (2002) Towards infinite-horizon optimality in nonlinear model predictive control. IEEE Trans Autom Control 47(4):679–682
20. Limon D, Alamo T, Camacho EF (2005) Enlarging the domain of attraction of MPC controllers. Automatica 41(2005):629–635
21. Rawlings JB, Mayne DQ (2009) Model predictive control: theory and design. Nob Hill Publishing. ISBN 9780975937709
22. Rawlings JB, Bonné D, Jørgensen JB, Venkat AN, Jørgensen SB (2008) Unreachable setpoints in model predictive control. IEEE Trans Autom Control 53(9):2209–2215
23. Domínguez-García AD, Hadjicostis CN (2011) Distributed algorithms for control of demand response and distributed energy resources. In: Proceedings of the 50th IEEE conference on decision and control and European control conference, (CDC-ECC). Orlando, Florida, pp 27–32

24. Garin F, Schenato L (2010) A survey on distributed estimation and control applications using linear consensus algorithms. Netw Control Syst 406:75–107
25. Maciejowski J (2002) Predictive control: with constraints. Pearson Education, New York

Chapter 5
Distributed Formation Control Using Population Games

Chapter 4 has presented an application of the distributed population dynamics to the design of a DMPC, i.e., as a complement in the design of the predictive controller to make it non-centralized. Nevertheless, these distribute dynamics have further applications. This chapter exploits the features of distributed population dynamics presented in Chap. 4 for a multi-agent system formation problem. More specifically, this chapter addresses the control for Unmanned Aerial Vehicles (UAVs). In general, different approaches are reported in the literature addressing the formation control of flying vehicles of diverse nature [1, 2], even considering non-centralized control topologies. Besides, some control approaches have been reported in [3–7]. The contributions presented in this chapter have been partially published in [8]. Additionally to what was discussed in [9], it is pointed out that the stability analysis developed in [9] is extensible to time-varying population structures. In this regard, it is suitable to implement the distributed population dynamics over engineering problems whose communication network varies along the time. In order to illustrate the powerful properties of distributed population dynamics, a distributed formation problem for multi-agent systems under time-varying communication network is proposed. To this end, it is assumed that each agent has a limited communication range, implying that each agent only has partial information about the entire set of agents. Furthermore, it is assumed that at least one follower agent is able to communicate to the leader agent that operates as a spatial reference for the whole formation. Additionally, it is shown that an agent can leave or enter the formation without having to modify controllers of the remaining agents. An additional relevant feature of the proposed distributed control strategy, the formation shape can be reconfigured dynamically along the time. Finally, simulation results present the behavior of the formation for two different scenarios. First, the case in which a new agent enters to the formation is tested. Secondly, the performance of the proposed distributed controller when the formation is modified along the time is shown.

© Springer International Publishing AG, part of Springer Nature 2019
J. Barreiro-Gomez, *The Role of Population Games in the Design
of Optimization-Based Controllers*, Springer Theses,
https://doi.org/10.1007/978-3-319-92204-1_5

5.1   UAV Model

The vehicles considered in this chapter are stabilized multicopters. It is assumed that each multicopter has an onboard controller (autopilot) that is capable of maintaining the vehicle in a hovering state and accepts generalized torque commands in the four independent degrees of freedom corresponding to position and yaw. The pose of the ith UAV is therefore described by $\mathbf{c}_i = [c_i^x \ \ c_i^y \ \ c_i^z \ \ c_i^\phi]^\top$. The adopted dynamic model of the internally stabilized ith vehicle can be described as

$$\mathbf{u}_i = \bar{\mathbf{A}}_i \ddot{\mathbf{c}}_i + \bar{\mathbf{B}}_i \dot{\mathbf{c}}_i, \tag{5.1}$$

where $\bar{\mathbf{A}}_i = \mathrm{diag}(\tilde{M}, \tilde{M}, \tilde{M}, I_z)$ is the generalized inertia matrix with vehicle mass $\tilde{M}$ and moment of inertia around the z axis I_z. Moreover, $\bar{\mathbf{B}}_i$ is the friction matrix, and $\mathbf{u}_i = [u_i^x \ \ u_i^y \ \ u_i^z \ \ u_i^\phi]^\top$ is the vector of generalized torques applied to the vehicle [5, 10]. These features are typically found in commercially-available multicopters and flight controllers [11]. Additionally, in Sect. 5.3.4 each UAV implements a local PID controller that actuates on the system described by (5.1) to track a desired input setpoint as shown in Fig. 5.2a. These models are implemented in the MATLAB/Simulink environment to generate the results of Sect. 5.3.6. From now on, due to the fact that the distributed control presented in this chapter is applicable for any multi-agent system, UAVs are going to be treated as agents in a multi-agent system.

Remark 5.1   Due to the fact that the formation control is known in the context of multi-agent systems, then (only in this chapter) UAVs are going to be treated as agents and individuals composing the population are going to be treated as decision makers.   ◇

5.2   Problem Statement Over Information Graphs

Consider a set of finite agents denoted by $\mathcal{A} = \{1, \ldots, n\}$, where $n \geq 2$. One of the agents is a leader denoted by $\ell \in \mathcal{A}$, which influences decisions of the entire set of the remainder agents known as followers, i.e., the set of follower agents is $\mathcal{F} = \mathcal{A} \backslash \{\ell\}$, and notice that $\mathcal{F} \neq \emptyset$. Each agent $i \in \mathcal{A}$ is located in the positive orthant space, i.e., at non-negative measurable coordinates denoted by c_i^x, c_i^y, and $c_i^z \in \mathbb{R}_{\geq 0}$; and with a measurable yaw angle denoted by $c_i^\phi \in \mathbb{R}_{\geq 0}$ that is considered to be also non-negative. All these measurements are collected in a vector denoted by $\mathbf{c}_i = [c_i^x \ \ c_i^y \ \ c_i^z \ \ c_i^\phi]^\top$, for all $i \in \mathcal{A}$. In addition, let $\mathcal{D} = \{x, y, z, \phi\}$ be the set of all possible position parameters for all the agents from $\mathcal{A}$.

On the other hand, it is assumed that each agent can communicate to all other agents, which are located within a communication-range radius denoted by $\psi_i \in \mathbb{R}_{>0}$, for all $i \in \mathcal{A}$. This situation leads to a time-varying communication network depending on the spatial distribution of agents. This communication network is represented by an undirected graph denoted by $\mathcal{G}(t) = (\mathcal{A}, \mathcal{E}(t), \mathbf{A}(t))$ (the graph is considered to be undirected since each link is assumed to be a bidirectional channel), where $\mathcal{A}$ is the set of nodes corresponding to each agent (i.e., the set of nodes in the graph is the same as the set of agents), $\mathcal{E}(t)$ is the set of links describing the communication network among agents, and $\mathbf{A}(t) = [a_{ij}(t)]$ is the adjacency matrix whose elements $a_{ij}(t) = 1$ if the ith agent shares information with the jth agent, and $a_{ij}(t) = 0$, otherwise. More precisely, the element $a_{ij}(t)$ depends on the communication range given by ψ_i as follows:

$$\xi_{ij} = \sqrt{(x_i - x_j)^2 + (y_i - y_j)^2 + (z_i - z_j)^2},$$
$$a_{ij}(t) = \max\left(0, \operatorname{sgn}\left[\psi_i - \xi_{ij}\right]\right),$$

where $a_{ij}(t) = a_{ji}(t)$. Then, the constructed adjacency matrix $\mathbf{A}(t)$ implies that each agent $i \in \mathcal{A}$ has a time-varying set of neighbors with whom it can communicate. The set of neighbors for the ith agent is given by $\mathcal{N}_i(t) = \{j : (i, j) \in \mathcal{E}(t)\}$.

Assumption 5.1 In order to achieve the leader tracking satisfying a desired formation shape in a distributed manner, the communication should be represented by a connected graph.   ◇

The leader agent $\ell \in \mathcal{A}$ goes over a pre-established trajectory, and constitutes a spatial reference for all the follower agents $i \in \mathcal{F}$ to perform a desired formation, i.e., the follower agents $i \in \mathcal{F}$ track the trajectory of the leader $\ell \in \mathcal{A}$ in an organized manner maintaining a required formation. Furthermore, notice that, depending on the engineering application, the interpretation of the leader might be changed by a target agent that is desired to track in an organized manner given a formation structure.

The objective formation is established by assigning desired Euclidean distances from each follower $i \in \mathcal{F}$ to the leader $\ell \in \mathcal{A}$ independently for each coordinate x, y, and z, and an angle for the yaw ϕ. These desired distances to achieve the formation are denoted by r_i^d, for all $i \in \mathcal{F}$, and $d \in \mathcal{D}$. Each agent has information about the formation reference distance represented by $\mathbf{r}_i = [r_i^x \; r_i^y \; r_i^z \; r_i^\phi]^\top$, for all $i \in \mathcal{F}$.

For instance, Fig. 5.1a shows a diagonal shape whose formation references are given by

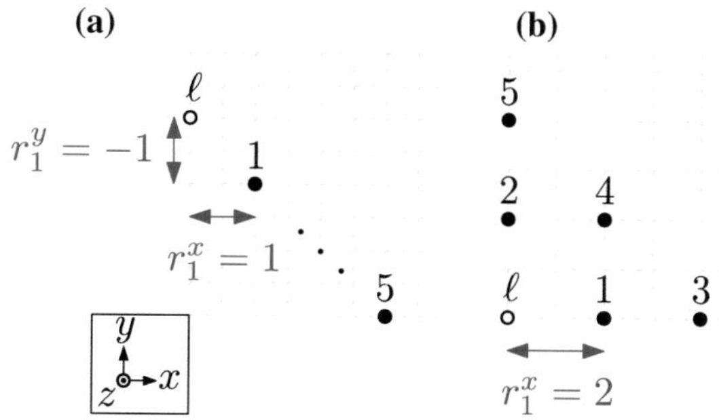

Fig. 5.1 Two different and possible formations with the same yaw angle ϕ_i, and the same z_i coordinate, for all the followers, and the leader. **a** diagonal formation, and **b** triangular formation (taken from [8])

$$\mathbf{r}_i = [i \quad -i \quad 0 \quad 0]^\top, \text{ for all } i \in \mathcal{F},$$

and Fig. 5.1b shows a triangular shape whose formation references are given by

$$\mathbf{r}_1 = [2 \ 0 \ 0 \ 0]^\top, \quad \mathbf{r}_2 = [0 \ 2 \ 0 \ 0]^\top,$$
$$\mathbf{r}_3 = [4 \ 0 \ 0 \ 0]^\top, \quad \mathbf{r}_4 = [2 \ 2 \ 0 \ 0]^\top,$$
$$\mathbf{r}_5 = [0 \ 4 \ 0 \ 0]^\top.$$

These formation references play an important role in the design of the distributed population-games-based formation controller presented in Sect. 5.3.

5.3 Population-Games Approach

Having stated the formation control problem, a distributed strategy based on population games is presented and its main properties and advantages are discussed. Preliminary concepts are introduced with subtle differences with respect to Chap. 4 due to the context of the application. The methodology to address the problem with this approach is presented, and detailed control schemes are explained. Also, an illustrative example for a linear formation is shown for clearness.

5.3.1 Preliminaries

Consider four different populations. Different from Chap. 4 where populations were composed of agents, in this chapter each population is composed of a large and finite number of rational decision makers. Each population represents a possible position parameter, i.e., the set of labels for the four populations is given by $\mathcal{D} = \{x, y, z, \phi\}$. Moreover, there is a set of n available strategies at each population denoted by $\mathcal{A} = \{1, \ldots, n\}$ for all $d \in \mathcal{D}$ (each strategy is associated to a UAV, which is an agent in this context). Then, decision makers decide to select a strategy from the set $\mathcal{A}$. Similarly as in Sect. 5.2, and for notation clearness, $\mathcal{F} = \mathcal{A}\backslash\{\ell\}$ that is also a subset of strategies, where $\ell \in \mathcal{A}$ is a fictitious strategy as in [12] related to the leader agent. Strategy ℓ is fictitious due to the fact that there is not a desired position and yaw rotation for the leader agent (leader UAV) since it has a pre-established trajectory.

The scalar value $p_i^d \in \mathbb{R}_{\geq 0}$ is a portion of decision makers selecting the strategy (agent/UAV) $i \in \mathcal{F}$ in the population (position parameter) $d \in \mathcal{D}$. Moreover, the portion of decision makers p_i^d is associated to the corresponding desired position of the ith agent in the position parameter $d \in \mathcal{D}$. In this regard, notice that the desired position and yaw rotation for the follower agents are given by

$\mathbf{p}_i = [p_i^x \; p_i^y \; p_i^x \; p_i^\phi]^\top \in \mathbb{R}_{\geq 0}^4$, for all $i \in \mathcal{F}$. Besides, there is a portion of decision makers selecting ℓ as in [12] denoted by $p_\ell^d \in \mathbb{R}_{\geq 0}$, where $\ell \in \mathcal{A}$, and $\ell \notin \mathcal{F}$.

On the other hand, $\mathbf{p}^d = [p_1^d \; \cdots \; p_n^d]^\top \in \mathbb{R}_{\geq 0}^n$ is a vector that collects all the portions of decision makers in the population $d \in \mathcal{D}$. This vector is known as the population state or the strategic distribution in the respective population. The set of possible strategic distributions for the population $d \in \mathcal{D}$ is given by the simplex

$$\Delta^d = \left\{ \mathbf{p}^d \in \mathbb{R}_{\geq 0}^n : \sum_{i \in \mathcal{A}} p_i^d = \pi^d \right\}, \tag{5.2}$$

where $\pi^d \in \mathbb{R}_{>0}$ denotes the population size. Furthermore, π^d is associated to the constrained space for the corresponding coordinates (x, y, or z) or rotation (ϕ) that agents should respect. Additionally, the interior of Δ^d is defined as follows:

$$\mathrm{int}\Delta^d = \left\{ \mathbf{p}^d \in \mathbb{R}_{>0}^n : \sum_{i \in \mathcal{A}} p_i^d = \pi^d \right\}.$$

Notice that the simplex in (5.2) implies that all the decision makers, corresponding with the follower agents, are constrained to be within a space characterized by

$$\sum_{i \in \mathcal{F}} p_i^d \leq \pi^d, \tag{5.3}$$

which means that it is desired that UAVs positions satisfy a constrained space

$$\begin{bmatrix} c_i^x \\ c_i^y \\ c_i^z \\ c_i^\phi \end{bmatrix} \leq \begin{bmatrix} \pi^x \\ \pi^y \\ \pi^z \\ \pi^\phi \end{bmatrix}, \; \forall \, i \in \mathcal{F}.$$

Decision makers in the population $d \in \mathcal{D}$ select among the different strategies in order to enhance their benefits. These benefits are described by a function denoted by f_i^d, and whose mapping is $f_i^d : \Delta^d \to \mathbb{R}$, for all $i \in \mathcal{A}$. Besides, let $\mathbf{f}^d$ be the vector of fitness functions whose mapping is given by $\mathbf{f}^d : \Delta^d \to \mathbb{R}^n$.

5.3.2 Distributed Replicator Dynamics and Their Properties

The distributed replicator dynamics are one of the fundamental distributed population dynamics presented in Sect. 4.2, which are given by

$$\dot{p}_i^d = p_i^d \left(f_i^d(\mathbf{p}^d) \sum_{j \in \mathcal{N}_i} p_j^d - \sum_{j \in \mathcal{N}_i} p_j^d f_j^d(\mathbf{p}^d) \right), \tag{5.4}$$

for all $i \in \mathcal{A}$, and $d \in \mathcal{D}$. Alternatively, (5.4) can be compacted as

$$\dot{\mathbf{p}}^d = \text{diag}(\mathbf{p}^d) \left[\text{diag}(\mathbf{f}^d(\mathbf{p}^d)) \mathbf{A} \mathbf{p}^d - \mathbf{A} \text{diag}(\mathbf{f}^d(\mathbf{p}^d)) \mathbf{p}^d \right], \tag{5.5}$$

for all $d \in \mathcal{D}$. The equilibrium point $\mathbf{p}^{d\star} \in \text{NE}(\mathbf{f}^d)$ of the distributed replicator dynamics equation (5.4) implies any of two situations, i.e., that the portion of decision makers is $p_i = 0$ for some $i \in \mathcal{A}$ (there is an extinction for a strategy), or that $f_i^d(\mathbf{p}^{d\star}) = f_j^d(\mathbf{p}^{d\star})$, for all $i, j \in \mathcal{A}, d \in \mathcal{D}$. Consequently, notice that if $\mathbf{p}^{d\star} \in \text{int}\Delta^d$, and if there is no extinction for any strategy, then the equilibrium point implies that $f_i^d(\mathbf{p}^{d\star}) = f_j^d(\mathbf{p}^{d\star})$, for all $i, j \in \mathcal{A}, d \in \mathcal{D}$.

Then, it is shown that the equilibrium point $\mathbf{p}^\star \in \text{int}\Delta^d$ is asymptotically stable under the distributed replicator dynamics.

Theorem 5.1 *Let $\mathbf{f}^d$ be a full-potential game with strictly concave potential function $V^d(\mathbf{p}^d)$, and let $\mathbf{p}^{d\star} \in \text{int}\Delta^d$ be a Nash equilibrium. If the graph $\mathcal{G} = (\mathcal{A}, \mathcal{E}, \mathbf{A})$ is connected and $\mathbf{p}^{d\star} \in \Delta^d$, then $\mathbf{p}^{d\star}$ is asymptotically stable under the distributed replicator dynamics (5.5).*

Proof See proof of Theorem 4.2.   □

In Corollary 5.1, it is highlighted that Theorem 5.1 also holds for time-varying communication sharing, broadening the spectrum in engineering applications to implement distributed population games as follows.

Corollary 5.1 *The asymptotic stability of $\mathbf{p}^{d\star} \in \text{int}\Delta^d$ under the distributed replicator dynamics (5.4) stated in Theorem 5.1 holds for connected time-varying graphs $\mathcal{G}(t) = (\mathcal{A}, \mathcal{E}(t), \mathbf{A}(t))$, i.e., for time-varying neighborhood $\mathcal{N}_i(t)$ and for all $i \in \mathcal{A}$. This statement can be seen from Theorem 5.1 since the Lyapunov function (4.6) is a common function for all possible connected-graph topologies.* ◇

Furthermore, the simplex Δ^d is an invariant set under the distributed replicator dynamics (5.4). This can be seen from the fact that (5.4) may be re-written as a function of the adjacency matrix, i.e.,

$$\dot{p}_i^d = \sum_{j \in \mathcal{A}} a_{ij} p_i^d p_j^d f_i^d(\mathbf{p}^d) - \sum_{j \in \mathcal{A}} a_{ij} p_j^d p_i^d f_j^d(\mathbf{p}^d), \tag{5.6}$$

for all $i \in \mathcal{A}$, and $d \in \mathcal{D}$. It follows that $\sum_{i \in \mathcal{A}} \dot{p}_i^d = 0$, for all $d \in \mathcal{D}$, due to the fact that $a_{ij} = a_{ji}$. This makes the first and second term in (5.6) equal, showing invariance property of the simplex Δ^d, for all $d \in \mathcal{D}$. This property also guarantees that the constraint in (5.3) is satisfied along the time.

5.3.3 Fitness Functions Design

Once the preliminary concepts of population games have been introduced with their modifications with respect to Chap. 4 and the distributed replicator dynamics have been presented, the design of the fitness functions is explained and discussed.

The distributed formation problem can be seen as a distributed control in charge of maintaining certain desired distances among a set of agents as presented in Sect. 5.2 (see Fig. 5.1). In this regard, if each agent is controlled in a manner that tracks the leader, then each agent maintains a constant distance to the leader agent, and the formation objective is achieved.

Taking advantage of the asymptotic convergence to a Nash equilibrium $\mathbf{p}^{d\star} \in \text{int}\Delta^d$ under the distributed replicator dynamics, the fitness functions are designed. Let $f_\ell^d = -c_\ell^d$ be the fitness function assigned to the portion of decision makers selecting the fictitious strategy associated to the leader agent $\ell \in \mathcal{A}$, $\ell \notin \mathcal{F}$. Now, let $f_i^d(p_i^d) = r_i^d - p_i^d$ be the fitness functions for all the follower agents $i \in \mathcal{F}$. First, notice that these fitness functions collected in $\mathbf{f}^d(\mathbf{p}^d)$ satisfy the condition for a stable game (see Definition 2.2). The matrix $D\mathbf{f}^d(\mathbf{p}^d)$ is a non-positive diagonal matrix, and with only one null element over the diagonal corresponding to the leader $\ell \in \mathcal{A}$.

Assuming that there is no extinction in the portion of decision makers, it is known that $\mathbf{p}^{d\star} \in \text{int}\Delta^d$ implies that $f_i^d(\mathbf{p}^{d\star}) = f_j^d(\mathbf{p}^{d\star})$, for all $i, j \in \mathcal{A}$, and $d \in \mathcal{D}$. Then,

$$f_\ell^d(\mathbf{p}^{d\star}) = f_i^d(\mathbf{p}^{d\star}),$$
$$-c_\ell^d = r_i^d - p_i^{d\star},$$
$$p_i^{d\star} = c_\ell^d + r_i^d, \tag{5.7}$$

for all $i \in \mathcal{F}$, and $d \in \mathcal{D}$. The equilibrium point condition in (5.7) shows that p_i^d, for all $i \in \mathcal{F}$, converges to values such that the formation references are met for all the coordinate positions and yaw angles (see Fig. 5.1). Besides, the portion of decision makers p_ℓ^d, for all $d \in \mathcal{D}$, and for the fictitious strategy, adopts a value that satisfies the simplex Δ^d.

Remark 5.2 Notice that it is not necessary that all the follower agents $i \in \mathcal{F}$ have communication to the leader agent $\ell \in \mathcal{A}$, $\ell \notin \mathcal{F}$ (this fact depends on the non-centralized communication topology given by $\mathcal{G}$). Nonetheless, all the follower agents –also those without communication with the leader– converge to the appropriate position to achieve the formation with a spatial reference given by the leader agent.
◇

5.3.4 Distributed Control Scheme

It is assumed that all the follower agents have an already designed local controller in charge of achieving a position set-point given by $\mathbf{p}_i = [p_i^x \quad p_i^y \quad p_i^x \quad p_i^\phi]^\top \in \mathbb{R}_{\geq 0}^4$

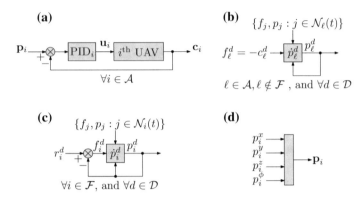

Fig. 5.2 Different control stages with their information dependence. **a** local PID position controller for each agent. **b** population-games-based formation control for the fictitious strategy representing the leader agent. **c** population-games-based formation control for all the follower agents. **d** collection of all the set-points for the local position controllers (taken from [8])

within the positive orthant space. In this chapter, these local controllers are PIDs as presented in Fig. 5.2a.

Moreover, depending on the current position of the agents, there is going to be a connected graph $\mathcal{G}(t)$ representing the communication network and possible information sharing. Then, the leader agent disposes of the information given by

$$\{(f_\ell, p_\ell)\} \cup \{(f_j, p_j) : j \in \mathcal{N}_\ell(t)\},$$

and the follower agents $i \in \mathcal{F}$ have information given by

$$\{(f_i, p_i)\} \cup \{(f_j, p_j) : j \in \mathcal{N}_i(t)\}.$$

Figure 5.2b shows the replicator-dynamics-based formation control for the fictitious strategy corresponding to the leader agent $\ell \in \mathcal{A}$, for all $d \in \mathcal{D}$. Figure 5.2c presents the replicator-dynamics-based formation control for all the follower agents $i \in \mathcal{F}$, and for all $d \in \mathcal{D}$. The formation control in Fig. 5.2c generates the appropriate position set-points, which are collected as presented in Fig. 5.2d, i.e., $\mathbf{p}_i = [p_i^x \ p_i^y \ p_i^x \ p_i^\phi]^\top$, which are established to the local controllers presented in Fig. 5.2a.

Then, the general control structure composed of local-position controllers for each agent, and the distributed formation control, constitutes a hierarchical scheme.

Figure 5.3 presents the overall hierarchical control scheme for the distributed replicator-dynamics-based formation controller with time-varying communication network. It can be seen that the leader agent has a local position controller whose reference describes a pre-established trajectory. On the other hand, all the follower agents have a local position controller whose position references denoted by $\mathbf{p}_i$, for all

Fig. 5.3 Hierarchical control scheme for the distributed multi-agent formation (taken from [8])

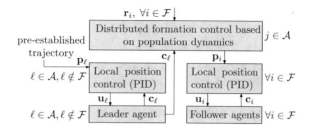

$i \in \mathcal{F}$, come from the upper layer, which is the distributed replicator-dynamics-based formation control. At this upper layer, all the position references $\mathbf{p}_i$, for all $i \in \mathcal{F}$, are computed in a distributed manner by using only information about the leader position $\mathbf{c}_i$ and the formation references $\mathbf{r}_i$, for all $i \in \mathcal{F}$. Notice that the replicator-dynamics-based formation control is distributed since each position reference is computed by using both local and partial information.

5.3.5 Communication Topology: Illustrative Example

An illustrative example to show the distributed communication dependency among $n = 6$ agents is presented.

Consider a control problem for the multi-agent formation presented in Fig. 5.1a. Moreover consider that, at certain time $t = \tilde{t}$, the communication graph is as follows: the set of nodes is $\mathcal{A} = \{1, \ldots, 6\}$, where the leader agent is $\ell = 1 \in \mathcal{A}$, and the follower agents are $\mathcal{F} = \{2, \ldots, 6\}$; and the set of links is given by $\mathcal{E}(\tilde{t}) = \{(1, 2), (2, 3), (3, 4), (4, 5), (5, 6)\}$. Then, the communication network, involving also the respective local controllers, is the one presented in Fig. 5.4.

Notice that follower agents $\{3, \ldots, 6\} \in \mathcal{F}$ can achieve a desired spatial position with respect to the leader agent even though they do not have information about its current position.

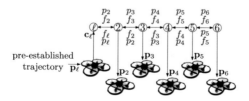

Fig. 5.4 Illustrative example for the distributed communication network in the replicator-dynamics-based formation control (taken from [8])

5.3.6 Population-Games-Based Control Results

In order to illustrate the performance of the population-games approach in the formation control, two different scenarios are considered for $n = 6$ agents. For these scenarios, and due to the fact that any agent from $\mathcal{A}$ may be the leader, it is selected arbitrarily to be $\ell = 6$. Scenarios are defined as follows:

Scenario 1: The first scenario consists in a linear formation (see Fig. 5.1a). Moreover, a follower agent $5 \in \mathcal{F}$ is initially located out of the communication range of all other agents. The leader trajectory is established such that it passes by certain coordinates, such that, along the time, the set of agents $\mathcal{A}$ can detect the isolated member, including it automatically in the formation. This scenario allows to illustrate the behavior of the distributed game-theoretical approach when a new agent is integrated into the problem.

Scenario 2: The second scenario consists in a linear formation (see Fig. 5.1a) that switches to the triangle formation (see Fig. 5.1b) along the time. This scenario illustrates the performance of the distributed game-theoretical approach for time-varying formation objectives.

Figure 5.5a shows the agents trajectory for the linear formation. Figure 5.5b, c show the evolution of the communication network. It can be seen that, at the beginning, the follower agent $5 \in \mathcal{F}$ is isolated and is not able to communicate with any other agent. Consequently, this agent remains without moving in the positive-orthant space until time $t = 138$ s, when it can communicate to agents three and four. Then, agent $5 \in \mathcal{F}$ is dynamically integrated to the formation problem. Furthermore, Fig. 5.5d–f show the evolution of the UAV positions c_i^d, for all $i \in \mathcal{A}$ and $d \in \mathcal{D}\backslash\{\phi\}$. It can be seen that follower agents maintain the desired distance to the leader in the coordinates x, and y, and that all the follower agents track the same position in z as the leader agent moves.

On the other hand, Fig. 5.6a shows the agents trajectory for the time-varying formation. First, a linear formation is established and, once this objective is achieved, the objective is changed to a triangular formation at time $t = 300$ s. Figure 5.6b, c show the evolution of the communication network. It can be seen that, when the formation is modified, the communication network is accommodated conveniently according to agents communication range. Furthermore, Fig. 5.6d–f show the evolution of the UAV positions c_i^d for all $i \in \mathcal{A}$ and $d \in \mathcal{D}\backslash\{\phi\}$. It can be seen that follower agents achieve the desired distances in all the position parameters r_i^d, for all $i \in \mathcal{F}$ and $d \in \mathcal{D}\backslash\{\phi\}$. Then, at time $t = 300$ s evolutions of the UAV positions vary to achieve the new formation objective.

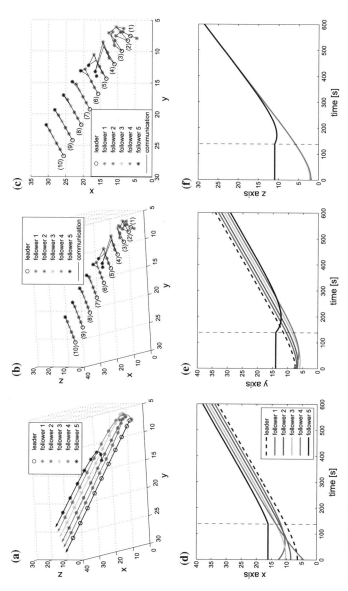

Fig. 5.5 Formation result for a line (see Fig. 5.1a), range communication $\psi_i = 3$ m, for all UAVs, and illustrating the modularity of the distributed replicator-dynamics-based formation control, i.e., incorporating a new agent to the formation problem along the time. **a** trajectories of the agents achieving the desired formation. **b, c** evolution of the communication graph for ten different time instants every 50 s, where sequences are as follows: (1) $t = 5$ s, (2) $t = 55$ s, ..., (10) $t = 455$ s. **d–f** evolution of the UAV positions c_i^d, for all UAVs, and position parameters; at $t = 138$ s the follower agent 5 is added to the formation problem (modified from [8])

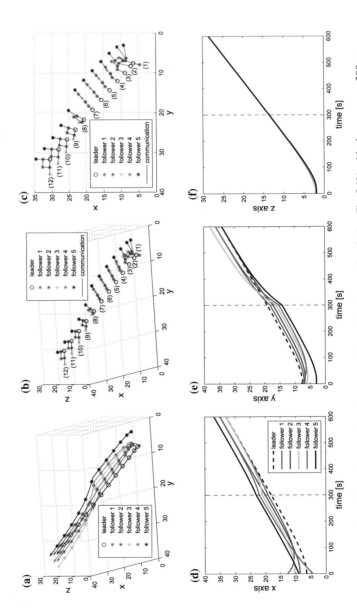

Fig. 5.6 Results for changing of topology along the time, from a line (see Fig. 5.1a) to a triangle (see Fig. 5.1b) with change at $t = 300$ s, range communication $\psi_i = 2.5$ m, for all UAVs. **a** trajectories of the agents achieving the desired formation for the line, and then for the triangle. **b, c** evolution of the communication graph for 12 different time instants every 50 s, where sequences are as follows: (1) $t = 5$ s, (2) $t = 55$ s, …, (12) $t = 555$ s. **d–f** evolution of the UAV positions c_i^d, for all UAVs, and position parameters; at $t = 300$ s the formation is changed (modified from [8])

5.4 Summary

A novel distributed formation control based on population games has been presented. It has been shown that the formation is achieved by using non-centralized communication structures among the agents. In addition, it has been shown that the proposed distributed controller can deal with time-varying formation objectives, and that the information sharing network can be varied conveniently as a function of the communication range for each agent. Under this time-varying graph scenario, it has been highlighted that stability remains as connectivity of the graph is preserved. Besides, a degree of modularity has been discussed, where a new agent can be incorporated or removed to/from the formation problem in a dynamical manner without affecting other agents controllers.

The fact that population games satisfy a single-coupled constraint has been used in this chapter for the design of a distributed controller. Moreover, the game theoretical approach presented in Chaps. 3 and 4, and in this one, cannot consider more than one coupled constraint, becoming a limitation for further engineering applications. Then, next chapter addresses the issue of considering multiple coupled constraints with a population-games approach.

References

1. Nazari M, Butcher E (2016) Decentralized consensus control of a rigid-body spacecraft formation with communication delay. J Guid Control Dyn 39(4):838–851
2. Zhou J, Hu Q (2013) Decentralized finite time attitude synchronization control of satellite formation flying. J Guid Control Dyn 36(1):185–195
3. Wang J, Xin M (2012) Integrated optimal formation control of multiple unmanned aerial vehicles. IEEE Trans Control Syst Technol 21:1731–1744
4. Kevinczky T, Borrelli F, Fregene K, Godbole D, Balas G (2008) Decentralized receiding horizon control and coordination of autonomous vehicle formation. IEEE Trans Control Syst Technol 16:19–33
5. Mas I, Kitts C (2014) Dynamic control of mobile multirobot systems: the cluster space formulation. IEEE Access 2:558–570
6. Mas I, Kitts C (2010) Centralized and decentralized multi-robot control methods using the cluster space control framework. In: IEEE/ASME international conference on advanced intelligent mechatronics (AIM). pp 115–122
7. Lin W (2014) Distributed UAV formation control using differential game approach. Aerosp Sci Technol 35:54–62
8. Barreiro-Gomez J, Mas I, Ocampo-Martinez C, Sánchez Peña R, Quijano N (2016) Distributed formation control of multiple unmanned aerial vehicles over time-varying graphs using population games. In: Proceedings of the 55th IEEE conference on decision and control (CDC). Las Vegas, USA, pp 5245–5250
9. Barreiro-Gomez J, Obando G, Quijano N (2017) Distributed population dynamics: optimization and control applications. IEEE Trans Syst Man Cybern Syst 47(2):304–314
10. Voos H, Nourghassemi B (2009) Nonlinear control of stabilized flight and landing for quadrotor UAVs. In: Proceedings of the 7th workshop on advanced control and diagnosis. Zielona Góra, Poland, pp 1–6

11. Lim H, Park J, Lee D, Kim HJ (2012) Build your own quadrotor: open-source projects on unmanned aerial vehicles. IEEE Robot Autom Mag 19(3):33–45
12. Pantoja A, Quijano N (2012) Distributed optimization using population dynamics with a local replicator equation. In: Proceedings of the 51st IEEE conference on decision and control (CDC). Maui, Hawaii, pp 3790–3795

Chapter 6
Distributed Predictive Control Using Density-Dependent Population Games

Chapters 3, 4 and 5 have presented the role of population games [1–6] in control applications involving only one coupled constraint. This chapter addresses the issue of dealing with multiple coupled constraints using a population-games approach that allows the population mass to vary along the time, i.e., considering *birth* and *death*. To this end, density games are studied. First, it is proposed to extend the mean dynamics with strategy-interaction constraints to the case considering a reproduction rate parameter, i.e., the density-dependent mean dynamics with non-complete population-interaction structures. Then, by using different revision protocols [1–3, 6], the distributed density-dependent replicator, Smith, and projection dynamics are deduced. Afterwards, it is shown that these density dynamics may be used to solve distributed constrained optimization problems when selecting properly the fitness functions in the strategic interaction based on a Lagrangian function. As a second part, and taking advantage of the properties that density games have, a DMPC controller design is proposed based on the distributed density-dependent population games (DDPG). Besides, the relationship between the population-interaction structure and the distributed information-sharing network for the DMPC controller is discussed. It is shown that the population-interaction structure can be modified dynamically along the time by adding conditions over the optimization problem constraints depending on the current system state, this fact leading to a time-varying information-sharing network for control purposes. Preliminary results corresponding to the contributions presented in this chapter have been published in [7].

6.1 Density-Dependent Population Games

The density-dependent mean dynamics are obtained from the mean dynamics (4.1) including a function of reproduction rate denoted by δ_i, for all $i \in \mathcal{S}$, i.e.,

© Springer International Publishing AG, part of Springer Nature 2019
J. Barreiro-Gomez, *The Role of Population Games in the Design of Optimization-Based Controllers*, Springer Theses,
https://doi.org/10.1007/978-3-319-92204-1_6

$$\dot{p}_i = \sum_{j \in \mathcal{N}_i} p_j \varrho_{ji}(\mathbf{f}(\mathbf{p}), \mathbf{p}) - p_i \sum_{j \in \mathcal{N}_i} \varrho_{ij}(\mathbf{f}(\mathbf{p}), \mathbf{p}) + \delta_i(\mathbf{f}(\mathbf{p}), \mathbf{p}), \ \forall i \in \mathcal{S}. \qquad (6.1)$$

When $\delta_i > 0$, then there is *birth* in the strategy $i \in \mathcal{S}$ since there are positive conditions promoting reproduction. On the other hand, if $\delta_i < 0$, then there is *death* (interpretation of negative reproduction rates) in the strategy $i \in \mathcal{S}$ due to the fact that there are not ideal conditions for reproduction. In this regard, δ_i should be directly proportional to the fitness function f_i, i.e., successful agents (those with greater fitness functions) have more chances to have offspring [8].

Definition 6.1 The reproduction rate, denoted by $\delta : \mathbb{R}^n \times \mathbb{R}^n_{\geq 0} \to \mathbb{R}^n$, is a function satisfying that reproduction rates decline as population increases [8], i.e., $\delta_i(\mathbf{f}(\mathbf{p}), \mathbf{p})$ should decrease as the portion p_i increases. $\diamond$

Remark 6.1 Notice that the population mass is not longer constant when considering the reproduction rate. Therefore, the possible population states are given by the set $\Delta^o = \{\mathbf{p} \in \mathbb{R}^n_{\geq 0}\}$, which is the positive orthant. $\diamond$

6.1.1 Distributed Density-Dependent Replicator Dynamics

First, the distributed density-dependent replicator dynamics (D3RD) are generated. In order to deduce a version of the D3RD, the pairwise proportional imitation protocol is used (see Table 2.1), i.e., $\varrho_{ij}(\mathbf{f}(\mathbf{p}), \mathbf{p}) = p_j [f_j(\mathbf{p}) - f_i(\mathbf{p})]_+$.

The interpretation of the revision protocol dictates that an agent selecting a strategy $i \in \mathcal{S}$, comparing itself with an agent selecting strategy $j \in \mathcal{S}$, decides to move to $j \in \mathcal{S}$ only if the change represents an improvement over its reproductive chances, i.e., if $f_j(\mathbf{p}) > f_i(\mathbf{p})$, and the switch rate is made in relation to the number of agents p_j selecting strategy $j \in \mathcal{S}$. Furthermore, the reproduction rate function is selected to be of the form $\delta_i(\mathbf{f}(\mathbf{p}), \mathbf{p}) = \beta_i f_i(\mathbf{p})$, where $\beta_i \geq 0$ can be either constant or time varying, e.g., $\beta_i = 1$, or $\beta_i = p_i$.

Under the framework of stable games, $f_i(p_i)$ is decreasing with respect to the number of agents p_i. The proposed reproduction-rate function is suitable according to Definition 6.1 since f_i declines as p_i increases, then δ_i also declines as p_i increases. Replacing $\varrho_{ij}(\mathbf{f}(\mathbf{p}), \mathbf{p})$, and $\delta_i(\mathbf{f}(\mathbf{p}), \mathbf{p})$ in (6.1) yields

$$\dot{p}_i = \sum_{j \in \mathcal{N}_i} p_j p_i [f_i(\mathbf{p}) - f_j(\mathbf{p})]_+ - p_i \sum_{j \in \mathcal{N}_i} p_j [f_j(\mathbf{p}) - f_i(\mathbf{p})]_+ + \beta_i f_i(\mathbf{p}), \ \forall i \in \mathcal{S},$$

where the first term is null when $f_j(\mathbf{p}) \geq f_i(\mathbf{p})$, and the second term is null when $f_i(\mathbf{p}) \geq f_j(\mathbf{p})$. Moreover, notice that the first term is the same as the second term with opposite sign. Taking these two observations into account, it follows that

$$\dot{p}_i = \sum_{j \in \mathcal{N}_i} p_j p_i \left(f_i(\mathbf{p}) - f_j(\mathbf{p}) \right) + \beta_i f_i(\mathbf{p}), \quad \forall i \in \mathcal{S}.$$

Finally, the D3RD are obtained as

$$\dot{p}_i = p_i \left(f_i(\mathbf{p}) \sum_{j \in \mathcal{N}_i} p_j - \sum_{j \in \mathcal{N}_i} p_j f_j(\mathbf{p}) \right) + \beta_i f_i(\mathbf{p}), \quad \forall i \in \mathcal{S},$$

or, in its matricial form depending on the adjacency matrix $\mathbf{A}$ of the graph $\mathcal{G}$ that describes the population-interaction structure, as

$$\dot{\mathbf{p}} = \text{diag}(\mathbf{p}) \left[\text{diag}\left(\mathbf{f}(\mathbf{p})\right) \mathbf{A}\mathbf{p} - \mathbf{A}\text{diag}\left(\mathbf{f}(\mathbf{p})\right) \mathbf{p} \right] + \text{diag}(\beta)\mathbf{f}(\mathbf{p}). \tag{6.2}$$

The equilibrium point of (6.2) is a Nash equilibrium denoted by $\mathbf{p}^\star \in \text{NE}(\mathbf{f})$, where $f_i(\mathbf{p}^\star) = f_j(\mathbf{p}^\star)$, for all $i, j \in \mathcal{S}$. Additionally, the equilibrium considering the reproduction rate implies that $\mathbf{f}(\mathbf{p}^\star) = \mathbf{0}_n$. The interpretation of this situation from a biological perspective is that there is no agent with incentives to move among strategies, since it could not increase its reproduction chances. Likewise, there is an equilibrium between birth and death ($\delta_i = 0$, for all $i \in \mathcal{S}$), such that the population size remains constant. Furthermore, the D3RD in (6.2) can be re-written simplifying the expression as in [9], i.e.,

$$\dot{\mathbf{p}} = \mathbf{L}^{(\mathbf{p})}\mathbf{f}(\mathbf{p}) + \text{diag}(\beta)\mathbf{f}(\mathbf{p}), \tag{6.3}$$

where $\beta \in \mathbb{R}^n_{\geq 0}$ is the vector of all the β_i for all $i \in \mathcal{S}$, and $\mathbf{L}^{(\mathbf{p})} = [l_{ij}^{(\mathbf{p})}]$ is a matrix depending on the population state $\mathbf{p}$, whose entries are defined as follows:

$$l_{ij}^{(\mathbf{p})} = \begin{cases} -a_{ij} p_i p_j, & \text{if } i \neq j \\ \displaystyle\sum_{r \in \mathcal{S}, r \neq i} a_{ir} p_i p_r, & \text{if } i = j. \end{cases}$$

6.1.2 Distributed Density-Dependent Smith Dynamics

The distributed density-dependent Smith dynamics (D3SD) are generated from the density-dependent mean dynamics in (6.1) and by using the pairwise comparison protocol (see Table 2.1), i.e., $\varrho_{ij}(\mathbf{f}(\mathbf{p}), \mathbf{p}) = \left[f_j(\mathbf{p}) - f_i(\mathbf{p}) \right]_+$.

The interpretation of the revision protocol indicates that an agent selecting a strategy $i \in \mathcal{S}$ switches to strategy $j \in \mathcal{S}$ if the change represents an improvement over its reproductive chances, i.e., if $f_j(\mathbf{p}) > f_i(\mathbf{p})$. Furthermore, the reproduction rate function is selected as in the case of the D3RD, i.e., $\delta_i(\mathbf{f}(\mathbf{p}), \mathbf{p}) = \beta_i f_i(\mathbf{p})$, where $\beta_i \geq 0$ can be either constant or time varying, e.g., $\beta_i = 1$, or $\beta_i = p_i$. Replacing $\varrho_{ij}(\mathbf{f}(\mathbf{p}), \mathbf{p})$, and $\delta_i(\mathbf{f}(\mathbf{p}), \mathbf{p})$ in (6.1) yields

$$\dot{p}_i = \sum_{j \in \mathcal{N}_i} p_j \left[f_i(\mathbf{p}) - f_j(\mathbf{p}) \right]_+ - p_i \sum_{j \in \mathcal{N}_i} \left[f_j(\mathbf{p}) - f_i(\mathbf{p}) \right]_+ + \beta_i f_i(\mathbf{p}), \ \forall i \in \mathcal{S},$$

which is the differential equation corresponding to the D3SD. Alternatively, the D3SD can be written as using simplification presented in [9], i.e.,

$$\dot{p}_i = \sum_{j \in \mathcal{N}_i} \frac{1}{2} \Big((1 - \nu_{ij}) p_i + (1 + \nu_{ij}) p_j \Big) \left[f_i(\mathbf{p}) - f_j(\mathbf{p}) \right] + \beta_i f_i(\mathbf{p}), \ \forall i \in \mathcal{S},$$

(6.4)

where $\nu_{ij} = \text{sgn}(f_i(\mathbf{p}) - f_j(\mathbf{p}))$. It follows that (6.4) can be expressed as

$$\dot{\mathbf{p}} = \tilde{\mathbf{L}}^{(\mathbf{p})} \mathbf{f}(\mathbf{p}) + \text{diag}(\boldsymbol{\beta}) \mathbf{f}(\mathbf{p}),$$

(6.5)

where $\tilde{\mathbf{L}}^{(\mathbf{p})} = [\tilde{l}_{ij}^{(\mathbf{p})}]$ is a matrix depending on the population state $\mathbf{p}$, whose entries are as follows:

$$\tilde{l}_{ij}^{(\mathbf{p})} = \begin{cases} -\dfrac{a_{ij}}{2} \Big((1 - \nu_{ij}) p_i + (1 + \nu_{ij}) p_j \Big), & \text{if } i \neq j \\[2ex] \displaystyle\sum_{r \in \mathcal{S}, r \neq i} \dfrac{a_{ir}}{2} \Big((1 - \nu_{ir}) p_i + (1 + \nu_{ir}) p_r \Big), & \text{if } i = j. \end{cases}$$

6.1.3 Distributed Density-Dependent Projection Dynamics

The distributed density-dependent Projection dynamics (D3PD) are deduced from the density-dependent mean dynamics in (6.1) and by considering the modified pairwise comparison protocol (see Table 2.1), i.e., $\varrho_{ij}(\mathbf{f}(\mathbf{p}), \mathbf{p}) = \dfrac{[f_j(\mathbf{p}) - f_i(\mathbf{p})]_+}{p_i}$.

Notice that the interpretation of the revision protocol is the same as the pairwise proportional imitation protocol with a different switching rate. In this case, the switching rate is proportional to the inverse of the portion of agents selecting strategy $i \in \mathcal{S}$. Moreover, the reproduction rate is selected to be the same as in the deduction of the D3RD. Replacing $\varrho_{ij}(\mathbf{f}(\mathbf{p}), \mathbf{p})$, and $\delta_i(\mathbf{f}(\mathbf{p}), \mathbf{p})$ in (6.1) yields

$$\dot{p}_i = \sum_{j \in \mathcal{N}_i} \left[f_i(\mathbf{p}) - f_j(\mathbf{p}) \right]_+ - \sum_{j \in \mathcal{N}_i} \left[f_j(\mathbf{p}) - f_i(\mathbf{p}) \right]_+ + \beta_i f_i(\mathbf{p}), \ \forall i \in \mathcal{S},$$

it follows that the D3PD are given by

$$\dot{p}_i = \sum_{j \in \mathcal{N}_i} [f_i(\mathbf{p}) - f_j(\mathbf{p})] + \beta_i f_i(\mathbf{p}), \ \forall i \in \mathcal{S}.$$

(6.6)

Finally, the D3PD can be re-written as

$$\dot{\mathbf{p}} = \mathbf{L}\mathbf{f}(\mathbf{p}) + \mathrm{diag}(\beta)\mathbf{f}(\mathbf{p}),$$

where $\mathbf{L} = [l_{ij}]$ is the Laplacian of the graph $\mathcal{G}$, i.e.,

$$l_{ij} = \begin{cases} -a_{ij}, & \text{if } i \neq j \\[2ex] \displaystyle\sum_{r \in \mathcal{S}, r \neq i} a_{ir}, & \text{if } i = j. \end{cases}$$

6.1.4 Stability Analysis

Once the density-dependent population dynamics have been formally introduced, a stability analysis of the equilibrium point is made. Then, it is shown that the mentioned Nash equilibrium $\mathbf{p}^\star \in \Delta^\circ$ is asymptotically stable under the density-dependent population dynamics with region of attraction given by the positive orthant as it is stated in the following theorems.

Theorem 6.1 *Let $\mathbf{f}$ be a full-potential game with strictly concave potential function $V(\mathbf{p})$, and let $\mathbf{p}^\star \in \Delta^\circ$ be a Nash equilibrium for a corresponding population size $\pi \in \mathbb{R}_{\geq 0}$ such that $\mathbf{f}(\mathbf{p}^\star) = \mathbf{0}_n$. If the population-interaction structure is given by a connected graph $\mathcal{G}$, then $\mathbf{p}^\star \in \Delta^\circ$ is asymptotically stable under the D3RD (6.2), and the D3SD (6.4).*

Proof Since $\mathbf{f}$ is a full-potential game with potential concave function $V(\mathbf{p})$, it is considered the same Lyapunov candidate function as in [9], i.e., $E_V(\mathbf{p}) = V(\mathbf{p}^\star) - V(\mathbf{p})$, where $E_V(\mathbf{p}^\star) = 0$, and $E_V(\mathbf{p}) > 0$, for all $\mathbf{p} \neq \mathbf{p}^\star$. Then, it follows that $\dot{E}_V(\mathbf{p}) = -(\nabla V(\mathbf{p}))^\top \dot{\mathbf{p}}$, which is the same as $\dot{E}_V(\mathbf{p}) = -\mathbf{f}(\mathbf{p})^\top \dot{\mathbf{p}}$. Now, replacing $\dot{\mathbf{p}}$ from (6.3) for the D3RD, it is obtained that

$$\begin{aligned} \dot{E}_V(\mathbf{p}) &= -\mathbf{f}(\mathbf{p})^\top \left(\mathbf{L}^{(\mathrm{p})}\mathbf{f}(\mathbf{p}) + \mathrm{diag}(\beta)\mathbf{f}(\mathbf{p}) \right), \\ &= -\mathbf{f}(\mathbf{p})^\top \mathbf{L}^{(\mathrm{p})}\mathbf{f}(\mathbf{p}) - \mathbf{f}(\mathbf{p})^\top \mathrm{diag}(\beta)\mathbf{f}(\mathbf{p}). \end{aligned} \tag{6.7}$$

For the first term in (6.7), notice that $\mathbf{L}^{(\mathrm{p})}$ corresponds to the Laplacian of a graph $\mathcal{G}^{(\mathrm{p})} = (\mathcal{S}, \mathcal{E}, \mathbf{A}^{(\mathrm{p})})$, where $\mathbf{A}^{(\mathrm{p})} = [a_{ij}^{(\mathrm{p})}]$ is the adjacency matrix with entries given by $a_{ij}^{(\mathrm{p})} = a_{ij} p_i p_j$. The entries of the adjacency matrix are non-negative due to the fact that $\mathbf{p} \in \Delta^\circ$, which is the positive orthant. Therefore $\mathbf{L}^{(\mathrm{p})} \succeq 0$ for any $\mathbf{p} \in \Delta^\circ$. Regarding the second term in (6.7), the diagonal matrix is $\mathrm{diag}(\beta) \succeq 0$ due to the fact that $\beta_i \geq 0$ for all $i \in \mathcal{S}$. Finally, it is concluded that $\dot{E}_V(\mathbf{p}) \leq 0$.

Regarding the D3SD, it is obtained that $\dot{E}_V(\mathbf{p}) = -\mathbf{f}(\mathbf{p})^\top \tilde{\mathbf{L}}^{(\mathrm{p})}\mathbf{f}(\mathbf{p}) - \mathbf{f}(\mathbf{p})^\top \mathrm{diag}(\beta)\mathbf{f}(\mathbf{p})$. The analysis is the same but with $\tilde{\mathbf{L}}^{(\mathrm{p})}$ corresponding to the Laplacian matrix of the graph $\mathcal{G}^{(\mathbf{p})} = (\mathcal{S}, \mathcal{E}, \mathbf{A}^{(\mathbf{p})})$ with $a_{ij}^{(\mathbf{p})} = (a_{ij}/2)((1 - \nu_{ij})p_i + (1 + \nu_{ij})p_j)$. Therefore, $\tilde{\mathbf{L}}^{(\mathbf{p})} \succeq 0$ for any $\mathbf{p} \in \Delta^\circ$, and it is concluded that $\dot{E}_V(\mathbf{p}) \leq 0$. The equality $\dot{E}_V(\mathbf{p}) = 0$ holds when $\mathbf{f}(\mathbf{p}) = \mathbf{0}_n$, and hence $\mathbf{p}^\star$ is asymptotically stable under the D3SD (6.4).

Moreover, $\dot{E}_V(\mathbf{p}) = 0$ in all the cases holds when $\mathbf{f}(\mathbf{p}) = \mathbf{0}_n$, and hence $\mathbf{p}^\star$ is asymptotically stable under the D3RD (6.2), and the D3SD (6.4) with region of attraction in the positive orthat Δ°. $\square$

Theorem 6.2 *Let $\mathbf{f}(\mathbf{p}) = \nabla V(\mathbf{p})$, where $V(\mathbf{p})$ is a strictly concave function, and let $\mathbf{p}^\star \in \mathbb{R}^n$ be an equilibrium point such that $\mathbf{f}(\mathbf{p}^\star) = \mathbf{0}_n$. If the population-interaction structure is given by a connected graph $\mathcal{G}$, then $\mathbf{p}^\star \in \mathbb{R}^n$ is globally asymptotically stable under the D3PD (6.6).*

Proof Consider the same Lyapunov candidate function as in Theorem 6.1, i.e., $E_V(\mathbf{p}) = V(\mathbf{p}^\star) - V(\mathbf{p})$, where $E_V(\mathbf{p}^\star) = 0$, and $E_V(\mathbf{p}) > 0$, for all $\mathbf{p} \neq \mathbf{p}^\star$. Then, it follows that $\dot{E}_V(\mathbf{p}) = - (\nabla V(\mathbf{p}))^\top \dot{\mathbf{p}}$, which is the same as $\dot{E}_V(\mathbf{p}) = -\mathbf{f}(\mathbf{p})^\top \dot{\mathbf{p}}$. Now, replacing $\dot{\mathbf{p}}$ from (6.6) for the D3PD, it is obtained that

$$\dot{E}_V(\mathbf{p}) = - \mathbf{f}(\mathbf{p})^\top (\mathbf{L}\mathbf{f}(\mathbf{p}) + \mathrm{diag}(\boldsymbol{\beta})\mathbf{f}(\mathbf{p})),$$
$$= - \mathbf{f}(\mathbf{p})^\top \mathbf{L}\mathbf{f}(\mathbf{p}) - \mathbf{f}(\mathbf{p})^\top \mathrm{diag}(\boldsymbol{\beta})\mathbf{f}(\mathbf{p}). \tag{6.8}$$

The first term in (6.8) is negative since $\mathbf{L} \succeq 0$ for a connected graph $\mathcal{G}$. Moreover, the second term in (6.8) is also negative since $\mathrm{diag}(\boldsymbol{\beta}) \succeq 0$ due to the fact that $\beta_i \geq 0$ for all $i \in \mathcal{S}$. Then, it is concluded that $\dot{E}_V(\mathbf{p}) \leq 0$. The equality $\dot{E}_V(\mathbf{p}) = 0$ holds when $\mathbf{f}(\mathbf{p}) = \mathbf{0}_n$, and therefore $\mathbf{p}^\star$ is globally asymptotically stable under the D3PD (6.6) with region of attraction in the $\mathbb{R}^n$ since $E_V(\mathbf{p})$ is radially unbounded. $\square$

Corollary 6.1 *The asymptotic stability of $\mathbf{p}^\star \in \Delta^\circ$ under the D3RD (6.2), and the D3SD (6.4) stated in Theorem 6.1; and $\mathbf{p}^\star \in \mathbb{R}^n$ under the D3PD (6.6) stated in Theorem 6.2 hold for connected time-varying graphs $\mathcal{G}(t) = (\mathcal{S}, \mathcal{E}(t), \mathbf{A}(t))$, i.e., for a time-varying neighborhood $\mathcal{N}_i(t)$, for all $i \in \mathcal{S}$. This statement is concluded since the postulated Lyapunov function $E_V(\mathbf{p}) = V(\mathbf{p}^\star) - V(\mathbf{p})$ is a common function for all possible connected-graph topologies.* $\diamond$

6.1.5   Full-Potential Game with DDPG

Consider the following constrained optimization problem as an illustrative example:

$$\underset{\mathbf{p}}{\text{maximize }} V(\mathbf{p}) = -(p_1 - 1)^2 - (p_2 - 2)^2, \tag{6.9a}$$

subject to

$$\sum_{i=1}^{2} p_i = \pi, \tag{6.9b}$$

$$p_1, p_2 \geq 0, \tag{6.9c}$$

where $\pi \in \mathbb{R}_{>0}$ is constant. The optimization problem (6.9) can be solved by using any of the six classical population dynamics, i.e., replicator, Smith, projection, Logit choice, BNN or best response dynamics [1]. To this end, function (6.9a) is taken as the potential function of a full-potential game (see Definition 2.3), i.e., $\mathbf{f}(\mathbf{p}) = \nabla V(\mathbf{p})$. Moreover, the initial condition for the classical population dynamics should satisfy the constraints (6.9b) and (6.9c), i.e., $\mathbf{p}(0) \in \mathbb{R}_{\geq 0}$, and $\sum_{i=1}^{2} p_i(0) = \pi$. Now, suppose that the constraint (6.9b) is no longer considered in the optimization problem, but (6.9c) is still imposed. Therefore, the new optimization problem can be solved with DDPG, i.e., the D3RD (6.2), the D3SD (6.4), or the D3PD (6.6), and with an initial condition belonging to the positive orthant $\mathbf{p}(0) \in \Delta^\circ$.

Figure 6.1 shows the phase plane for both classical population dynamics, and density-dependent population dynamics. It can be seen that the classical population dynamics converge to the optimal point subject to constraint (6.9b) with $\pi = p_1(0) + p_2(0)$. In contrast, the density-dependent population dynamics converge to the optimal point without considering the constraint (6.9b), i.e., $\mathbf{p}^\star = [1 \quad 2]^\top \in \Delta^\circ$ for any initial condition in the positive orthant $\mathbf{p}(0) \in \Delta^\circ$. The solution coincides for both approaches as long as $\pi = 3$ for the classical population dynamics.

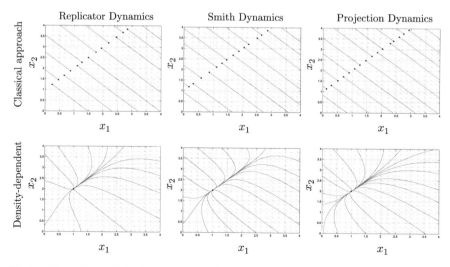

Fig. 6.1 Phase plane of trajectories corresponding to the replicator, Smith and projection dynamics with the full-potential game $\mathbf{f}(\mathbf{p})$ for both classical approach and density-dependent approach. Black dots show the equilibrium points for initial conditions in the positive orthant

6.1.6 Solving Constrained Optimization Problems with DDPG

Consider a quadratic programming (QP) optimization problem of the form

$$\underset{\mathbf{y}}{\text{maximize}} \ \ f(\mathbf{y}), \tag{6.10a}$$

$$\text{subject to} \quad \mathbf{Ey} \leq \mathbf{e}, \tag{6.10b}$$

$$\mathbf{Gy} = \mathbf{g}, \tag{6.10c}$$

$$\mathbf{y} \in \mathbb{R}^{v}_{\geq 0}, \tag{6.10d}$$

where $f : \mathbb{R}^{v}_{\geq 0} \to \mathbb{R}$ is concave, and continuously differentiable. Moreover, $\mathbf{E} \in \mathbb{R}^{q \times v}$, and $\mathbf{e} \in \mathbb{R}^{q}$ allow to state the q inequality constraints (6.10b), and $\mathbf{G} \in \mathbb{R}^{r \times v}$, and $\mathbf{g} \in \mathbb{R}^{r}$ define the r equality constraints (6.10c).

Inequality constraints can be transformed into equality constraints by adding non-negative slack variables denoted by $\mathbf{s} \in \mathbb{R}^{q}_{\geq 0}$. To this end, consider the vector variable $\boldsymbol{\xi} = [\mathbf{y}^{\top} \ \mathbf{s}^{\top}]^{\top} \in \mathbb{R}^{p}$, where $p = v + q$, then the QP optimization problem (6.10) is reformulated as follows:

$$\underset{\boldsymbol{\xi}}{\text{maximize}} \ \ f(\boldsymbol{\xi}), \tag{6.11a}$$

$$\text{subject to} \quad \mathbf{H}\boldsymbol{\xi} = \mathbf{h}, \tag{6.11b}$$

$$\boldsymbol{\xi} \in \mathbb{R}^{p}_{\geq 0}, \tag{6.11c}$$

where $f : \mathbb{R}^{p}_{\geq 0} \to \mathbb{R}$ is concave, and continuously differentiable. The matrix $\mathbf{H} \in \mathbb{R}^{w \times p}$, and $\mathbf{h} \in \mathbb{R}^{w}$ determine the w equality constraints (6.11b), where $w = q + r$.

Now, omitting the positiveness constraints (6.11c), the Lagrangian function $L : \mathbb{R}^{p} \times \mathbb{R}^{w} \to \mathbb{R}$ is [10]

$$L(\boldsymbol{\xi}, \boldsymbol{\mu}) = f(\boldsymbol{\xi}) + \boldsymbol{\mu}^{\top} (\mathbf{H}\boldsymbol{\xi} - \mathbf{h}), \tag{6.12}$$

where $\boldsymbol{\mu} \in \mathbb{R}^{w}$ corresponds to the Lagrange multipliers associated to the w equality constraints of (6.11). Moreover, $\nabla_{\boldsymbol{\xi}} L(\boldsymbol{\xi}, \boldsymbol{\mu}) = \nabla f(\boldsymbol{\xi}) + \mathbf{H}^{\top}\boldsymbol{\mu}$, and $-\nabla_{\boldsymbol{\mu}} L(\boldsymbol{\xi}, \boldsymbol{\mu}) = -\mathbf{H}\boldsymbol{\xi} + \mathbf{h}$.

The Lagrange condition is used to find the possible extreme points $\boldsymbol{\xi}^{\star} \in \mathbb{R}^{p}$ of the function $f(\boldsymbol{\xi})$ (maximum of the function f) subject to constraints (6.11b), in which

$$\begin{bmatrix} \nabla_{\boldsymbol{\xi}} L(\boldsymbol{\xi}^{\star}, \boldsymbol{\mu}^{\star}) \\ -\nabla_{\boldsymbol{\mu}} L(\boldsymbol{\xi}^{\star}, \boldsymbol{\mu}^{\star}) \end{bmatrix} = \mathbf{0}_{p+w}. \tag{6.13}$$

Now, let $\mathbf{p} = [\boldsymbol{\xi}^{\top} \ \boldsymbol{\mu}^{\top}]^{\top} \in \mathbb{R}^{n}$ be the vector representing the number of agents in a strategic interaction with $\mathcal{S} = \{1, \dots, n\}$, where $n = p + w$. Besides, let

$$\mathbf{f}(\mathbf{p}) = [\nabla_\xi L(\xi, \mu)^\top \quad -\nabla_\mu L(\xi, \mu)^\top]^\top \tag{6.14}$$

be the fitness functions corresponding to all the strategies $\mathcal{S}$. The population game (6.14) can be seen as two different potential games $\mathbf{f}_\xi(\xi, \mu) = \nabla_\xi L(\xi, \mu)$, and $\mathbf{f}_\mu(\xi, \mu) = -\nabla_\mu L(\xi, \mu)$, whose potential functions are $L(\xi, \mu)$, and $-L(\xi, \mu)$, respectively. Therefore, notice that the Hessian of the potential functions is $\nabla_\xi^2 L(\xi, \mu) = \nabla^2 f(\xi)$, and $\nabla_\mu^2(-L(\xi, \mu)) = \mathbf{0}$. Therefore, $\mathbf{f}_\xi(\xi, \mu)$, and $\mathbf{f}_\mu(\xi, \mu)$ are stable games [1]. Finally, since $\mathbf{f}(\mathbf{p})$ is a full-potential and stable game, and according to Theorem 6.1 and Corollary 6.1, the optimization problem (6.10) can be solved in a distributed way by using the D3RD (6.2), the D3SD (6.4), or the D3PD (6.6), and under time-varying graphs $\mathcal{G}(t)$.

Remark 6.2 Even though population dynamics only admit values within the positive orthant Δ°, an optimization problem allowing negative values can also been solved by using the D3RD (6.2), the D3SD (6.4), or the D3PD (6.6). This is made by applying a change of variables in the fitness functions, e.g., if it is necessary to consider a constraint of the form $\underline{\mathbf{y}} \leq \mathbf{y} \leq \bar{\mathbf{y}}$, and let $\underline{\mathbf{y}} < \mathbf{0}$ and $\bar{\mathbf{y}} > \mathbf{0}$ be a lower and upper bound, respectively, it is possible to make a change of variables as $\mathbf{0} \leq \mathbf{y} - \underline{\mathbf{y}} \leq \bar{\mathbf{y}} - \underline{\mathbf{y}}$, or equivalently, $\mathbf{0} \leq \tilde{\mathbf{y}} \leq \bar{\mathbf{y}} - \underline{\mathbf{y}}$. Obtaining an optimization problem expressed in the required form as in (6.10). $\diamond$

Remark 6.2 shows that the proposed method with DDPG is versatile to solve optimization problems with different types of inequality constraints in a distributed manner. As an application, the next section presents a DMPC controller design based on the D3RD (6.2), the D3SD (6.4), or the D3PD (6.6).

Proof-of-Concept Example

Let $\mathcal{S} = \{1, \ldots, 7\}$ be the set of strategies in a population with restricted information sharing information given by an undirected connected graph $\mathcal{G}$, which is presented in Fig. 6.2a. Consider the following constrained optimization problem:

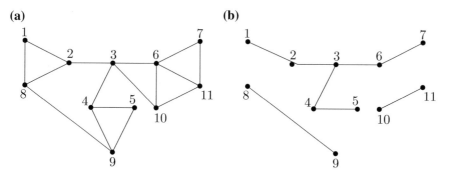

Fig. 6.2 Communication network for a population with seven strategies. **a** graph to compute the fitness functions, and **b** graph separating full-potential games

$$\underset{\mathbf{p}}{\text{maximize}} \quad -\frac{1}{2}\mathbf{p}^\top\mathbf{p} + [25 \quad 23 \quad 20 \quad 21 \quad 24 \quad 19 \quad 26]\mathbf{p}, \tag{6.15a}$$

subject to

$$p_i \geq 0, \quad \forall i \in \mathcal{S}, \tag{6.15b}$$

$$\begin{pmatrix} 50 \\ 16 \\ 25 \\ 18 \end{pmatrix} = \begin{pmatrix} 1 & 1 & 0 & 0 & 0 & 0 & 0 \\ 0 & 0 & 0 & 1 & 1 & 0 & 0 \\ 0 & 0 & 1 & 0 & 0 & 1 & 0 \\ 0 & 0 & 0 & 0 & 0 & 1 & 1 \end{pmatrix}\mathbf{p}. \tag{6.15c}$$

In order to solve the problem with DDPG, it is necessary to compute the Lagrangian function and determine the fitness functions as in (6.14). Due to the fact that the Theorems 6.1 and 6.2 hold for full-potential games, then $\mathbf{f_p}(\mathbf{p}, \mu) = \nabla_{\mathbf{p}}L(\mathbf{p}, \mu)$, and $\mathbf{f_\mu}(\mathbf{p}, \mu) = -\nabla_\mu L(\mathbf{p}, \mu)$ should be solved independently, even though those games share information in order to compute the fitness functions, i.e., fitness functions are computed using the graph shown in Fig. 6.2a. It is desired that all games converge to an equilibrium point such that $\mathbf{f_p}(\mathbf{p}^\star, \mu^\star) = \mathbf{0}$, and $\mathbf{f_\mu}(\mathbf{p}^\star, \mu^\star) = \mathbf{0}$, i.e., the same fitness function. Therefore, games can be solved independently. In addition, it is possible to solve different groups of games as it is illustrated next, i.e., the game is solved by using the information-sharing graph presented in graph Fig. 6.2b corresponding to three independent full-potential games. Also, those games may be solved by using different density-dependent population dynamics.

In order to solve the optimization problem with the Lagrangian function, there are dynamics associated to the Lagrange multipliers. Moreover, it is possible that the Lagrange multipliers get negative values. This fact represents a problem for the D3RD and for the D3SD since those dynamics can only evolve in the positive orthant Δ° (see Theorem 6.1). Therefore, to solve the optimization problem using the D3RD

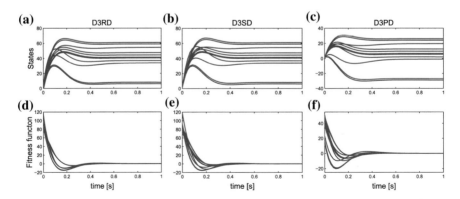

Fig. 6.3 Evolution of the population states **a**–**c**, and fitness functions **d**–**f** with DDPG for the example (6.15). Sub-figures **a** and **d** D3RD with an offset of 35, **b** and **e** D3SD with an offset of 35, and **c** and **f** D3PD

and the D3SD, a change of variable must be made to establish an *offset*. On the other hand, the fact that the Lagrange multipliers can get negative values is not an inconvenient for the D3PD since trajectories evolve in $\mathbb{R}^n$ (see Theorem 6.2).

Figure 6.3 shows the evolution of the proportion of agents under the D3RD with *offset*, the D3SD with *offset*, and the D3PD for the game corresponding to the optimization problem (6.15). It can be seen that the fitness functions converge to zero and proportion of agents to the solution.

6.2 DMPC Controller Based on DDPG with Time-Varying Information-Sharing Network

Consider the state-space discrete-time system presented in (3.1) with a sampling time τ, i.e.,

$$\mathbf{x}_{k+1} = \mathbf{A}_d \mathbf{x}_k + \mathbf{B}_d \mathbf{u}_k + \mathbf{B}_l \mathbf{d}_k, \tag{6.16}$$

where vectors $\mathbf{x} \in \mathbb{R}^{n_x}$, $\mathbf{u} \in \mathbb{R}^{n_u}$, and $\mathbf{d} \in \mathbb{R}^{n_d}$ denote the states, control inputs and disturbances, respectively as it has been presented in Chap. 3, and the state-space matrices are given by $\mathbf{A}_d \in \mathbb{R}^{n_x \times n_x}$, $\mathbf{B}_d \in \mathbb{R}^{n_x \times n_u}$, and $\mathbf{B}_l \in \mathbb{R}^{n_x \times n_d}$. States and control inputs are subject to physical and operational constraints that define feasible sets denoted by $\mathcal{X} \triangleq \{\mathbf{x} \in \mathbb{R}^{n_x} : \underline{\mathbf{x}} \leq \mathbf{x} \leq \bar{\mathbf{x}}\}$, and $\mathcal{U} \triangleq \{\mathbf{u} \in \mathbb{R}^{n_u} : \underline{\mathbf{u}} \leq \mathbf{u} \leq \bar{\mathbf{u}}\}$, where vectors $\underline{\mathbf{x}}$ and $\bar{\mathbf{x}}$ correspond to the lower and upper limits for the system states, respectively. Similarly, vectors $\underline{\mathbf{u}}$ and $\bar{\mathbf{u}}$ denote the lower and upper limits for the control inputs, respectively. The control sequence for a fixed-time $H_p \in \mathbb{Z}_{>0}$ at time instant $k \in \mathbb{Z}_{\geq 0}$ is denoted by $\hat{\mathbf{u}}_k$ (see (3.3a)). When the control-input sequence $\hat{\mathbf{u}}_k$ is applied to the system (6.16) with initial state $\mathbf{x}_{k|k} \triangleq \mathbf{x}_k$, a system states sequence $\hat{\mathbf{x}}_k$ is generated (see (3.3b)). Finally, the time-varying sequence of system disturbances along H_p is denoted by $\hat{\mathbf{d}}_k$ (see (3.3c)).

The system in (6.16) is controlled with an MPC controller whose optimization problem can be formulated as follows:

$$\underset{\mathbf{U}_k}{\text{minimize}} \quad \mathbf{U}_k^\top \mathbf{\Phi} \mathbf{U}_k + \boldsymbol{\phi}_k^\top \mathbf{U}_k, \tag{6.17a}$$

$$\text{subject to} \quad \mathbf{E} \mathbf{U}_k \leq \mathbf{e}_k, \tag{6.17b}$$

$$\mathbf{G} \mathbf{U}_k = \mathbf{g}_k. \tag{6.17c}$$

Moreover, for the optimization problem (6.17) vectors $\mathbf{e}_k$ and $\mathbf{g}_k$, which determine the inequality and equality constraints, vary every discrete-time instant k.

Remark 6.3 The optimization problem behind the MPC controller considers $q = 2n_u + 2n_x$ inequality constraints corresponding to the physical and operational limits for the system states and control, i.e., $\underline{\mathbf{x}} \leq \mathbf{x} \leq \bar{\mathbf{x}}$ and $\underline{\mathbf{u}} \leq \mathbf{u} \leq \bar{\mathbf{u}}$, respectively. In this regard, it is assumed that the constraints are organized in the QP problem (6.17) as follows: the first n_u constraints in (6.17b) are associated to the upper limits for the control inputs $\bar{\mathbf{u}}$, the next n_u constraints to the lower limits for the control inputs $\underline{\mathbf{u}}$, the next n_x constraints to the upper limits for the states $\bar{\mathbf{x}}$, and finally, the last n_x constraints to the lower limits for the states $\underline{\mathbf{x}}$. ◇

Therefore, notice that the QP problem formulation in (6.17) for the MPC controller has the same form as the optimization problem in (6.10). Furthermore, by adding non-negative slack variables $\mathbf{s} \in \mathbb{R}_{\geq 0}^q$, the cost function may be re-written, and the constraints can be compacted, i.e.,

$$\underset{\xi}{\text{minimize}} \underbrace{\begin{bmatrix} \mathbf{U}_k^\top & \mathbf{s}^\top \end{bmatrix}}_{\xi_k^\top} \underbrace{\begin{bmatrix} \boldsymbol{\Phi} & \mathbf{0}_{n_u \times q} \\ \mathbf{0}_{q \times n_u} & \mathbf{0}_{q \times q} \end{bmatrix}}_{\boldsymbol{\Psi}} \underbrace{\begin{bmatrix} \mathbf{U}_k \\ \mathbf{s} \end{bmatrix}}_{\xi_k} + \underbrace{\begin{bmatrix} \boldsymbol{\phi}_k^\top & \mathbf{0}_q^\top \end{bmatrix}}_{\psi_k^\top} \underbrace{\begin{bmatrix} \mathbf{U}_k \\ \mathbf{s} \end{bmatrix}}_{\xi_k}, \qquad (6.18a)$$

subject to

$$\underbrace{\begin{bmatrix} \mathbf{E} & \mathbb{I}_q \\ \mathbf{G} & \mathbf{0}_{r \times q} \end{bmatrix}}_{\mathbf{H}} \underbrace{\begin{bmatrix} \mathbf{U}_k \\ \mathbf{s} \end{bmatrix}}_{\xi_k} = \underbrace{\begin{bmatrix} \mathbf{e}_k \\ \mathbf{g}_k \end{bmatrix}}_{\mathbf{h}_k}. \qquad (6.18b)$$

Having added the slack variables, the optimization problem behind the MPC controller is formulated in the form (6.11), and it can be solved in a distributed manner by using the D3RD (6.2), the D3SD (6.4), or the D3PD (6.6), as explained in Sect. 6.1.

Assumption 6.1 States are measurable and their current value are available in order to state the QP problem in (6.17). ◇

Regarding the information dependence, it is mainly given by the fitness functions coupling. In order to determine the information-sharing structure for the D3RD (6.2), the D3SD (6.4), or the D3PD (6.6), the fitness functions (6.14) are expressed in the form

$$\mathbf{f}(\mathbf{p}) = \begin{bmatrix} f_1(\mathbf{p}) \\ \vdots \\ f_n(\mathbf{p}) \end{bmatrix} = \begin{bmatrix} \mathbf{f}_1(\mathbf{p}) \\ \mathbf{f}_2(\mathbf{p}) \\ \mathbf{f}_3(\mathbf{p}) \end{bmatrix}. \qquad (6.19)$$

The first block of the vector $\mathbf{f}(\mathbf{p})$ in (6.19) corresponding to the n_u control input variables is given by

$$\mathbf{f}_1(\mathbf{p}) = \begin{bmatrix} \dfrac{\partial L(\mathbf{p})}{\partial p_1} & \cdots & \dfrac{\partial L(\mathbf{p})}{\partial p_{n_u}} \end{bmatrix}^\top,$$

while its second block corresponding to the q inequality constraints is given by

$$\mathbf{f}_2(\mathbf{p}) = -\begin{bmatrix} \dfrac{\partial L(\mathbf{p})}{\partial p_{(n_u+1)}} & \cdots & \dfrac{\partial L(\mathbf{p})}{\partial p_{(n_u+q)}} \end{bmatrix}^\top.$$

Finally, the third block corresponds to the r equality constraints satisfying the order according to Remark 6.3, i.e.,

$$\mathbf{f}_3(\mathbf{p}) = -\left[\frac{\partial L(\mathbf{p})}{\partial p_{(n_u+q+1)}} \quad \cdots \quad \frac{\partial L(\mathbf{p})}{\partial p_{(n_u+q+r)}}\right]^{\top}.$$

Assumption 6.2 In the population game $\mathbf{f}$, convergence to the Nash equilibrium under the D3RD (6.2), the D3SD (6.4), or the D3PD (6.6), is achieved in shorter time than the sampling time τ for the discrete system in (6.16). ◇

Notice that the satisfaction of Assumption 6.2 depends on the type of system under control, i.e., the required time to compute the distributed population dynamics should be evaluated in comparison to how fast the dynamical system is. For instance, the case study proposed in this chapter has a sampling time of an hour, for which the Assumption 6.2 is satisfied.

In order to check from which strategies it is necessary to get information, the Hessian matrix of the Lagrangian function $\Theta = [\theta_{ij}] = \nabla^2 L(\mathbf{p})$ is computed, i.e., $\theta_{ij} = \frac{\partial f_i(\mathbf{p})}{\partial p_j}$. Then, the biggest required information-sharing matrix denoted by $\tilde{\Theta} = [\tilde{\theta}_{ij}]$, which corresponds to the existing coupling among the portion of agents at each strategy, is given by

$$\tilde{\theta}_{ij} = \begin{cases} 1, & \text{if } \theta_{ij} \neq 0, \\ 0, & \text{otherwise.} \end{cases} \tag{6.20}$$

For the scenario with constant graph $\mathcal{G}$, the adjacency matrix is given by the biggest required information-sharing matrix, i.e., $\mathbf{A} = \tilde{\Theta}$. Nevertheless, conditions over the adjacency matrix $\mathbf{A}$ can be added in order to use less information-sharing links when convenient. It is highlighted that conditions can be versatile and the time-varying graph can be addressed in different ways besides the one proposed in this chapter.

It is proposed to have an information-sharing graph topology depending on the necessary active constraints. This is made since, under some system state conditions, it is not necessary to consider the whole set of constraints. In this regard, some of them can be properly neglected reducing both the size of the information-sharing network and the computational burden. Figure 6.4 shows different possible regions for states and control inputs established by parameters $\bar{g}_i^x$, $\underline{g}_i^x$, $\bar{g}_i^u$, and $\underline{g}_i^u$, which are constant values determined at a design stage.

Regions describe *non-safe* sectors in the feasible set, or sectors near limits of a constraint. Figure 6.4 presents two examples, i.e., for a time $k < k_1$, the constraint $x_i \geq \underline{x}_i$ is active whereas the constraints $x_i \leq \bar{x}_i$ is neglected. Then, for the time $k > k_2$, the constraint $x_i \geq \underline{x}_i$ is neglected whereas the constraints $x_i \leq \bar{x}_i$ is considered. Similarly, Fig. 6.4 also shows the scenario of control input u_j with the respective time instants k_3, and k_4. The regions, shown in Fig. 6.4, are formally defined as follows:

- Upper region for states: $\bar{R}_i^x = \{x_i : \bar{g}_i^x \leq x_i \leq \bar{x}_i\}$,

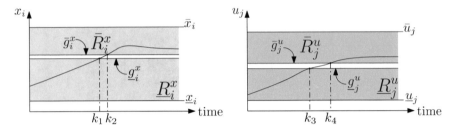

Fig. 6.4 Different regions for the consideration of inequality constraints

- Lower region for states: $\underline{R}_i^x = \{x_i : \underline{x}_i \le x_i \le \underline{g}_i^x\}$,
- Upper region for inputs: $\bar{R}_j^u = \{u_j : \bar{g}_j^u \le u_j \le \bar{u}_j\}$,
- Lower region for inputs: $\underline{R}_j^u = \{u_j : \underline{u}_j \le u_j \le \underline{g}_j^u\}$,

for all $i = 1, \ldots, n_x$, and $j = 1, \ldots, n_u$. Besides, binary variables $\bar{\gamma}_{i,k}^x, \underline{\gamma}_{i,k}^x$, which indicate whether or not the current state $x_{i,k}$ belongs to a region, are defined as follows:

$$\bar{\gamma}_{i,k}^x = \begin{cases} 1, & \text{if } x_{i,k} \in \bar{R}_i^x \\ 0, & \text{otherwise,} \end{cases}$$

$$\underline{\gamma}_{i,k}^x = \begin{cases} 1, & \text{if } x_{i,k} \in \underline{R}_i^x \\ 0, & \text{otherwise,} \end{cases}$$

where $i = 1, \ldots, n_x$. Parameters $\bar{\gamma}_{j,k}^u$, and $\underline{\gamma}_{j,k}^u$, indicating whether or not the current control input $u_{i,k}$ belongs to a given region, are stated similarly for $j = 1, \ldots, n_u$. These binary variables lead to a vector that determines the active and non-active constraints for states and control inputs at each time instant, i.e.,

$$\mathbf{\Gamma}_k^u = \left[\bar{\gamma}_{1,k}^u, \ldots, \bar{\gamma}_{n_u,k}^u, \underline{\gamma}_{1,k}^u, \ldots, \underline{\gamma}_{n_u,k}^u\right]^\top,$$

$$\mathbf{\Gamma}_k^x = \left[\bar{\gamma}_{1,k}^x, \ldots, \bar{\gamma}_{n_x,k}^x, \underline{\gamma}_{1,k}^x, \ldots, \underline{\gamma}_{n_x,k}^u\right]^\top.$$

Then, let $\tilde{\mathbf{\Gamma}}_k$ be the diagonal matrix of the active constraints at time instant $k \in \mathbb{Z}_{\ge 0}$ in the same order as explained in Remark 6.3, i.e.,

$$\tilde{\mathbf{\Gamma}}_k = \text{diag}\left(\left[\mathbf{1}_{n_u}^\top \quad \mathbf{\Gamma}_k^{u\top} \quad \mathbf{\Gamma}_k^{x\top} \quad \mathbf{1}_r^\top\right]\right).$$

Finally, these conditions over the active constraints lead to a time-varying graph with adjacency matrix $\mathbf{A}(t)$ that varies its topology every τ, given by $\mathbf{A}(k\tau) = \tilde{\mathbf{\Gamma}}_k\mathbf{\Theta}\tilde{\mathbf{\Gamma}}_k$, and the topology $\mathbf{A}(k\tau)$ is maintained during a time τ, i.e., $\mathbf{A}(t) = \mathbf{A}(k\tau)$,

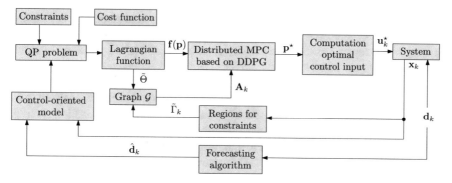

Fig. 6.5 Summary of the DMPC controller with distributed DDPG

for all $k\tau \le t < (k+1)\tau$. Alternatively, time-varying adjacency matrix can be denoted that $\mathbf{A}_k = \tilde{\boldsymbol{\Gamma}}_k \tilde{\boldsymbol{\Theta}} \tilde{\boldsymbol{\Gamma}}_k$.

Figure 6.5 shows the scheme corresponding to the DMPC controller based on DDPG with time-varying information-sharing network. The *non-safe* regions are determined in function of the current system state. Therefore, the information-sharing graph is established with an adjacency matrix $\mathbf{A}_k$. Finally, the DMPC controller based on DDPG computes the optimal control input at each time instant k satisfying the information-sharing restrictions.

6.3 Case Study: Barcelona Water Supply Network

Consider the same case study presented in Sect. 3.3.3, which is the aggregate model of the BWSN presented in Fig. 3.4. This benchmark has been studied, for instance, in [2, 7, 11–13]. An MPC controller is designed for the BWSN considering the slew rate $\Delta\mathbf{u}_{k+j} = \mathbf{u}_{k+j} - \mathbf{u}_{k+j-1}$, for all $j \in [0, H_p - 1] \cap \mathbb{Z}_{>0}$. Therefore, the corresponding optimization problem behind the MPC controller is as follows:

$$\underset{\mathbf{u}_{k|k},\ldots,\mathbf{u}_{k+H_p|k}}{\text{minimize}} \; J(\mathbf{x}, \mathbf{u}) = \sum_{j=1}^{H_p} \left(\|\mathbf{x}_{k+j|k} - \mathbf{x}_r\|_{\tilde{\mathbf{Q}}} + \|\Delta\mathbf{u}_{k+j|k}\|_{\tilde{\mathbf{R}}} + \gamma |\left(\alpha_1 + \alpha_{2,k+j}\right)^{\top}\mathbf{u}_{k+j|k}| \right),$$

$$(6.21a)$$

subject to

$$\mathbf{x}_{k+j+1|k} = \mathbf{A}_d\mathbf{x}_{k+j|k} + \mathbf{B}_d\mathbf{u}_{k+j|k} + \mathbf{B}_l\mathbf{d}_{k+j|k}, \quad \forall j \in [0, H_p - 1] \cap \mathbb{Z}_{\ge 0}, \quad (6.21b)$$

$$\mathbf{u}_{k+j|k} \in \mathcal{U}, \quad \forall j \in [0, H_p - 1] \cap \mathbb{Z}_{\ge 0}, \quad (6.21c)$$

$$\mathbf{x}_{k+j|k} \in \mathcal{X}, \quad \forall j \in [1, H_p] \cap \mathbb{Z}_{\ge 0}, \quad (6.21d)$$

$$\mathbf{0}_r = \mathbf{E}_u\mathbf{u}_{k+j|k} + \mathbf{E}_d\mathbf{d}_{k+j|k}, \quad \forall j \in [0, H_p - 1] \cap \mathbb{Z}_{\ge 0}, \quad (6.21e)$$

where $\mathbf{x}_r \in \mathbb{R}^{n_x}$ is a constant desired set-point for the system states $\mathbf{x} \in \mathbb{R}^{n_x}$. More-over, $\alpha_1 \in \mathbb{R}^{n_u}$ represents the time-invariant costs associated to the water resource, and $\alpha_2 \in \mathbb{R}^{n_u}$ represents the time-varying costs associated to the operation of valves and pumps. On the other hand, $\mathbf{E}_u \in \mathbb{R}^{r \times n_u}$, and $\mathbf{E}_d \in \mathbb{R}^{r \times n_d}$ construct the r equal-ity constraints in (6.21e). The matrices $\tilde{\mathbf{Q}} \in \mathbb{R}^{n_x \times n_x}$ and $\tilde{\mathbf{R}} \in \mathbb{R}^{n_u \times n_u}$, and the scalar $\gamma \in \mathbb{R}_{\geq 0}$ are weights assigning a prioritization for the control objectives related to the error and to energy slew rate, respectively. Assuming that the optimization problem (6.21) is feasible, an optimal sequence is computed, and following the MPC philoso-phy, a new optimization problem is formulated for the next time instant [14]. Notice that the optimization problem in (6.21) is also suitable for the real application since it minimizes an error with respect to a desired volume that may vary along the time, and it also takes into account the smoothness operation in order to avoid damage in the network. Finally, the economical aspect to minimize the costs associated to water and energy are also included in the cost function.

Then, some re-formulations over the cost function (6.21a) and constraints (6.21c)–(6.21e) are presented to obtain a QP problem. These modifications are necessary in order to show the density-dependent population-games approach as an alternative tool for DMPC controller design. The optimization problem behind the MPC controller in (6.21) can be conveniently re-formulated with a cost function given by

$$J = (\mathbf{X}_k - \mathbf{X}_r)^\top \mathbf{Q}\,(\mathbf{X}_k - \mathbf{X}_r) + \Delta\mathbf{U}_k^\top \mathbf{R}\Delta\mathbf{U}_k + \gamma\mathbf{U}_k^\top\alpha,$$

where weighting matrices are: $\mathbf{Q} = \mathrm{diag}([\tilde{\mathbf{Q}} \ \ldots \ \tilde{\mathbf{Q}}])$, $\mathbf{R} = \mathrm{diag}([\tilde{\mathbf{R}} \ \ldots \ \tilde{\mathbf{R}}])$, and

$$\alpha = \left[\left(\alpha_1 + \alpha_{2,k+1}\right)^\top \ \ \ldots \ \ \left(\alpha_1 + \alpha_{2,k+H_p}\right)^\top \right]^\top.$$

The reference vector along H_p is $\mathbf{X}_r = [\mathbf{x}_r^\top \ \mathbf{x}_r^\top \ \ldots \ \mathbf{x}_r^\top]^\top$, and vectors $\mathbf{X}_k$ and $\mathbf{U}_k$ are composed of the same elements of sequences in (3.3a) and (3.3b), i.e.,

$$\mathbf{X}_k = [\mathbf{x}_{k+1|k}^\top \ \mathbf{x}_{k+2|k}^\top \ \ldots \ \mathbf{x}_{k+H_p|k}^\top]^\top,$$
$$\mathbf{U}_k = [\mathbf{u}_{k|k}^\top \ \mathbf{u}_{k+1|k}^\top \ \ldots \ \mathbf{u}_{k+H_p-1|k}^\top]^\top.$$

Moreover, applying the appropriate transformations [14], the optimization prob-lem can be re-written as a QP problem in (6.17). Moreover, the BWSN is controlled under two different scenarios, i.e.,

Scenario 1: DMPC controller based on DDPG with time-varying information-sharing network for the BWSN.

Scenario 2: CMPC controller with constant information-sharing network for the BWSN.

In order to evaluate the performance of the control strategy, two different KPIs are proposed. One of them associated to the economical costs for operating the actuators as presented in (3.15), and the other one associated to the communication costs depending on the information-sharing links that are required to perform each control scheme. These KPIs are defined as follows:

- *Economical costs*: these costs correspond to the required energy, and the time-varying water costs during a day, i.e.,

$$\text{KPI}_{\text{Ecosts}}(\text{day}) = \sum_{k=1+24(\text{day}-1)}^{24+24(\text{day}-1)} (\alpha_1 + \alpha_{2,k})^\top \mathbf{u}_k. \tag{6.22}$$

- *Communication costs*: these costs correspond to the required permanent information-sharing links to compute the control inputs during a day, i.e.,

$$\text{KPI}_{\text{Ccosts}}(\text{day}) = \sum_{k=1+24(\text{day}-1)}^{24+24(\text{day}-1)} \frac{(\mathbb{1}_n^\top \mathbf{M}_k \mathbb{1}_n)}{2}, \tag{6.23}$$

where $\mathbf{M}_k = \mathbf{A}_k$ for the control strategy proposed in this chapter. The KPI in (6.23) is also used later in this thesis by modifying the matrix $\mathbf{M}_k$.

For the simulation results, the reference has been selected to be $\mathbf{x}_r = 0.6\bar{\mathbf{x}}$, and the weights in the cost function are selected to be $\tilde{\mathbf{Q}} = \mathbb{I}_{n_x}$, $\tilde{\mathbf{R}} = 1000\mathbb{I}_{n_u}$, and $\gamma = 1$. Furthermore, the regions to determine the activeness of constraints are computed by using the following parameters: $\bar{g}_i^x = \underline{g}_i^x = 0.6\bar{x}_i$, for all $i = 1, \ldots, n_x$; and $\bar{g}_j^u = 0.65\bar{u}_j$, and $\underline{g}_j^u = 0.35\bar{u}_j$, for all $j = 1, \ldots, n_u$.

Figure 6.6a presents the information-sharing topology required to solve the optimization problem by using the D3RD (6.2), the D3SD (6.4), or the D3PD (6.6), when all the inequality constraints are active, i.e., $\mathbf{A}(t) = \mathbf{\Theta}$. This topology also corresponds to the topology at $k = 1$ to initialize the control strategy when considering time-varying graphs. Figure 6.6b corresponds to the graph when adopting time-varying graphs and the proposed distributed density-dependent population-dynamics-based DMPC controller. Graph in Fig. 6.6d has a reduction of 36.15% of the information-sharing links. This reduction in the number of links from Fig. 6.6a to d is produced due to the fact that, at $k = 77$, inequality constraints are non-active according to the regions presented in Fig. 6.4. Furthermore, Fig. 6.6b and c correspond to time instants $k = 33$ and $k = 11$, respectively.

Figure 6.7 shows the evolution of some states achieving the imposed reference, also reflecting a proper performance of the proposed distributed density-dependent population-dynamics-based DMPC achieving the references and minimizing abrupt changes in the control signals –it is quite important to highlight that the evolution of system states and control inputs are exactly the same for both Scenario 1 and 2–. On the other hand, Fig. 6.7 also presents the behavior of some control signals. It can be seen that these control inputs oscillate in order to satisfy the demands. That is why these control inputs have the same periodicity as the disturbances (period of 24 hours). Figure 6.9 shows the evolution of the number of connected links in the information-sharing network along the time. It can be seen that, at the beginning,

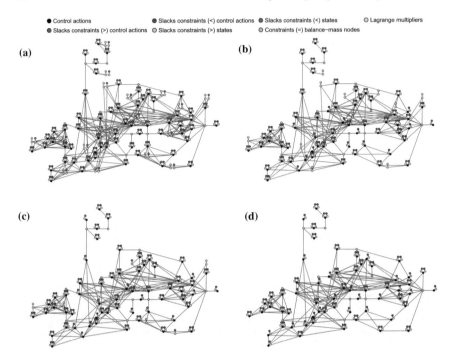

Fig. 6.6 Communication topologies for the DMPC controller based on DDPG. **a** information-sharing topology for the DMPC controller based on DDPG, i.e., $\mathbf{A} = \tilde{\boldsymbol{\Theta}}$. The number of links in this graph for this case study is 614; and time-varying information-sharing topology for the DMPC controller based on DDPG at time instant: **b** $k = 33$, i.e., $\mathbf{A}(33\tau) = \tilde{\boldsymbol{\Gamma}}(33)\tilde{\boldsymbol{\Theta}}\tilde{\boldsymbol{\Gamma}}(33)$, **c** $k = 11$, i.e., $\mathbf{A}(11\tau) = \tilde{\boldsymbol{\Gamma}}(11)\tilde{\boldsymbol{\Theta}}\tilde{\boldsymbol{\Gamma}}(11)$, and **d** $k = 77$, i.e., $\mathbf{A}(77\tau) = \tilde{\boldsymbol{\Gamma}}(77)\tilde{\boldsymbol{\Theta}}\tilde{\boldsymbol{\Gamma}}(77)$. The number of links in the graphs for this case study are: **b** 468, **c** 435, and **d** 392

it is needed to have the information-sharing graph corresponding to the biggest required information-sharing matrix, i.e., $\mathbf{A}(t) = \tilde{\boldsymbol{\Theta}}$. Then, after few iterations, the system reduces the number of required links considerably. Moreover, it can be seen a periodic behavior of the number of required links in the information-sharing network, is daily (period of 24 hours) as the disturbances, i.e., although all the demands have different magnitudes and mean values, they have the same daily periodicity as the disturbances d_4, d_8, d_{16}, and d_{24} presented in Fig. 6.8.

Table 6.1 presents the economical and communication KPIs corresponding to both considered scenarios. It can be seen that the economical costs associated to the water and the required energy to operate the valves and pumps are the same for both scenarios, i.e., the evolution of control inputs are equivalent even though the information-sharing networks are different. This fact regarding the difference of information-sharing networks is evidenced in the KPI in (6.23) (compare the total cost corresponding to $\mathrm{KPI}_{\mathrm{Ccost}}$ for both scenarios in Table 6.1). Moreover, it can be

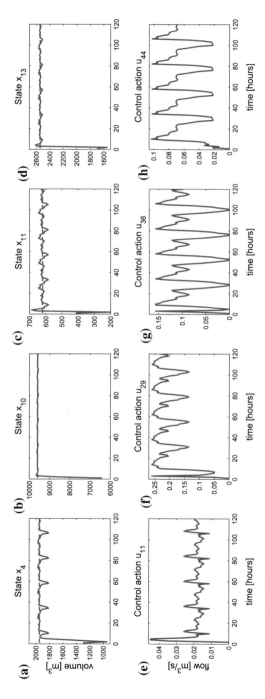

Fig. 6.7 The evolution of four states are presented at first row, i.e., x_4, x_{10}, x_{11}, and x_{13}. The evolution of four control inputs are presented at second row, i.e., u_{11}, u_{29}, u_{36}, and u_{44}

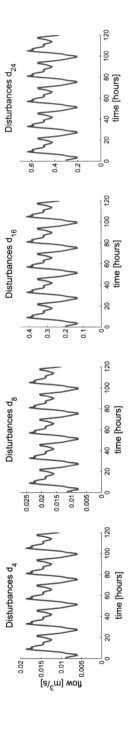

Fig. 6.8 The evolution of four demand profiles, i.e., disturbances d_4, d_8, d_{16}, and d_{24}

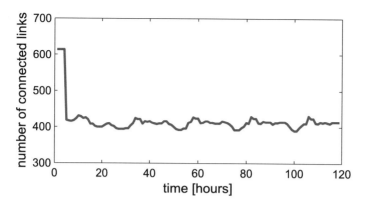

Fig. 6.9 Number of connected information-sharing links along the time. The minimum achieved number of links is 392 corresponding to Fig. 6.6d

Table 6.1 Summary of KPIs corresponding to operation of actuators and communication links

Day	Scenario 1		Scenario 2	
	KPI_{Ecosts} (e.u.)	KPI_{Ccost} (e.u.)	KPI_{Ecosts} (e.u.)	KPI_{Ccost} (e.u.)
1	24.2456	10724	24.2456	14736
2	21.6782	9800	21.6782	14736
3	21.4634	9828	21.4634	14736
4	21.3294	9838	21.3294	14736
5	21.2401	9840	21.2401	14736
Total	109.9567	50030	109.9567	73680
Overall*	50139.95		73789.65	

*The overall cost is computed by adding the total costs for both KPIs, i.e., the total values for $KPI_{Ecosts} + KPI_{Ccost}$

seen that there is a reduction in the communication costs for the second scenario, and therefore, there is a reduction in the overall costs without affecting the performance of the closed-loop system.

6.4   Summary

A general methodology to generate distributed density-dependent population dynamics has been presented by considering a reproduction rate in the distributed mean dynamics. Furthermore, the relationship between the equilibrium point of density games with the optimal point has been shown in a constrained optimization problem by selecting the description of benefits throughout the strategies using the Lagrangian of a potential function. Besides, the asymptotic stability of the equilibrium point under the D3RD has been formally proven for constant and time-varying population structures. Then, after introducing this class of dynamics and their properties, they have been applied to the design of a DMPC controller under a time-varying communication network. Simulation results have shown a reduction in the number of links in the information-sharing network over a 36% with respect to the total number of links for the communication network without neglecting any constraint.

References

1. Sandholm WH (2010) Population games and evolutionary dynamics. MIT Press, Cambridge
2. Barreiro-Gomez J, Quijano N, Ocampo-Martinez C (2016a) Constrained distributed optimization: a population dynamics approach. Automatica 69:101–116
3. Quijano N, Ocampo-Martinez C, Barreiro-Gomez J, Obando G, Pantoja A, Mojica-Nava E (2017) The role of population games and evolutionary dynamics in distributed control systems. IEEE Control Syst 37(1):70–97
4. Zino L, Como G, Fagnani F (2017) On imitation dynamics in potential population games. In: Proceedings of the 56th IEEE conference on decision and control (CDC). Melbourne, Australia, pp 757–762
5. Poveda JI, Brown PN, Marden JR, Teel AR (2017) A class of distributed adaptive pricing mechanisms for societal systems with limited information. In: Proceedings of the 56th IEEE conference on decision and control (CDC). Melbourne, Australia, pp 1490–1495
6. Lasaulce S, Tembine H (2011) Game theory and learning for wireless networks: fundamentals and applications. Academic Press, London
7. Barreiro-Gomez J, Quijano N, Ocampo-Martinez C (2016b) Distributed MPC with time-varying communication network: a density-dependent population games approach. In: Proceedings of the 55th IEEE conference on decision and control (CDC). Las Vegas, USA, pp 6068–6073
8. Novak S, Chatterjee K, Nowak MA (2013) Density games. J Theor Biol 334(2013):26–34
9. Barreiro-Gomez J, Obando G, Quijano N (2017a) Distributed population dynamics: optimization and control applications. IEEE Trans Syst Man Cybern Syst 47(2):304–314
10. Chong EKP, Zak SH (2013) An introduction to optimization. Wiley series in discrete mathematics and optimization. Wiley, New York
11. Barreiro-Gomez J, Ocampo-Martinez C, Quijano N (2017) Dynamical tuning for mpc using population games: a water supply network application. ISA Trans 69(2017):175–186
12. Barreiro-Gomez J, Ocampo-Martinez C, Quijano N (2017c) Partitioning for large-scale systems: a sequential dmpc design. In: Proceedings of the 20th IFAC world congress. Toulouse, France, pp 8838–8843
13. Barreiro-Gomez J, Ocampo-Martinez C, Quijano N (2015) Evolutionary-game-based dynamical tuning for multi-objective model predictive control. In: Olaru S, Grancharova A, Lobo Pereira F (eds) Developments in model-based optimization and control. Springer, Verlag, pp 115–138
14. Maciejowski J (2002) Predictive control: with constraints. Pearson education, New York

Chapter 7
Power Index in Control

Previous chapters have studied the role of non-cooperative-game approaches in the design of controllers. In contrast, this chapter discusses in studying a cooperative-game approach. Indices of power are alternative ways to solve a game, which are characterized for satisfying a certain system of axioms. Some of these power indices are, among others, the Shapley value, the Banzhaf–Coleman index, or the dictatorial index [1]. Moreover, even though for some examples different indices are equivalent, in the general case they adopt different values. Specifically, this chapter addresses the role of the Shapley power index in the design of controllers. In [1], it has been shown that the computation of the Shapley value is more complex than the computation of other power indices. The main reason is that the computation of the Shapley value involves a combinatorial explosion since it evaluates all the possible coalitions that can be made among players. Therefore, this fact implies a high computational burden. In addition, the computation requires information from all the players, i.e., it is computed under centralized communication structures, which constitutes an additional challenge for real applications, specially in cases with large amount of players, where big communication networks would be needed.

The issue of computing the Shapley power index under distributed structures is discussed. This chapter proposes a different way to compute the Shapley value for a specific characteristic function in the cooperative game in order to reduce the computational burden, which is one of the main issues when using this game-theoretical approach. Unlike the classical cooperative-games approach, which has a limited application due to the combinatorial explosion issues, the alternative method allows calculating the Shapley value in polynomial time and hence can be applied to large-scale problems (games involving large amount of players).

In addition, most of game-theoretical contributions address engineering problems from either a cooperative or non-cooperative perspective. However, this chapter shows that there must be control problems involving both cooperation and competition. Therefore, it is suitable to solve it by using both directions. In this chapter, the Shapley value and also population games are used together to solve a unique multi-objective control problem.

© Springer International Publishing AG, part of Springer Nature 2019 133
J. Barreiro-Gomez, *The Role of Population Games in the Design
of Optimization-Based Controllers*, Springer Theses,
https://doi.org/10.1007/978-3-319-92204-1_7

In order to show the performance of the proposed methodology based on both cooperative and non-cooperative games, a resource allocation problem in a water system treated in [2] is presented, and the stability of the closed-loop system is analyzed by using passivity theory as in [3, 4]. Moreover, analysis regarding invariant-set properties that guarantee a limitation-resource constraint and the stability of the whole closed-loop system are presented extending preliminary result presented in [5]. The resource allocation problem has been addressed with both replicator and projection dynamics since they share some important gradient characteristics, which have been studied in [6]. Besides, it is discussed the fact that the main result related to the cooperative-game approach allows to compute the Shapley value under a distributed structure, i.e., without considering a complete graph connecting the whole set of players. Also, it is shown that this result leads to an additional alternative to compute the Shapley value through a linear set of equations and using the coalitional rationality axiom (in a centralized manner, or under a distributed structure). As a consequence of the decrement of required communication links to make the computation under distributed structures, there is an increment in the computational cost. However, a comparison between three alternatives to compute the Shapley value, including the computation under a distributed structure, has been presented in function of the number of players. This comparison shows that the computation under a distributed structure is faster with respect to the traditional computation of the Shapley value as there is an increment in the number of players.

7.1 Problem Statement in Distribution Flow-Based Networks

Several engineering problems may be addressed as a flow-based distribution network just by defining general elements such as suppliers and demands, flow capacity, and storage capacity [7]. Consider a specific resource flow-based distribution network, which is composed only of three types of elements:

- **Storage**: element that stores the resource and with both inflows and outflows. Let $\mathcal{V} = \{1, \ldots, n\}$ be the set of $n \in \mathbb{Z}_{>0}$ storage nodes.[1]
- **Sink**: element that receives the resource from a storage node and with only inflows. Let $\mathcal{B} = \{n + 1, \ldots, 2n\}$ be the set $n \in \mathbb{Z}_{>0}$ of sink nodes.
- **Source**: element that provides the resource and with only outflows. Let $\mathcal{R} = \{2n + 1\}$ be the singleton set of a source node.

---

[1]The set $\mathcal{V}$ has three different interpretations depending on the context in which it is used, i.e.,

(a) $\mathcal{V}$ is the set of n storage nodes in the distribution flow-based networks context (Sect. 7.1).
(b) $\mathcal{V}$ is the set of n strategies in the population-games context (Sect. 7.2).
(c) $\mathcal{V}$ is the set of n players in the cooperative-games context (Sect. 7.3).

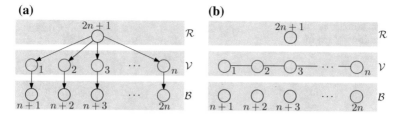

Fig. 7.1 Distribution flow-based network. **a** directed graph corresponding to the existing network flows. **b** undirected-graph example corresponding to the communication sharing among sub-systems (associated to storage nodes)

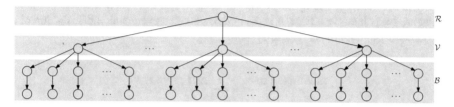

Fig. 7.2 Tree-shaped topology composed by multiple flow-based distribution networks

Consider a directed graph denoted by $\tilde{\mathcal{G}} = (\tilde{\mathcal{V}}, \tilde{\mathcal{E}})$, where $\tilde{\mathcal{V}} = \{\mathcal{V} \cup \mathcal{B} \cup \mathcal{R}\}$ is the set of $2n + 1$ nodes representing the storage, sink, and source elements in the flow-based distribution network, and $\tilde{\mathcal{E}} \subset \{(r, \ell) : r, \ell \in \tilde{\mathcal{V}}\}$ is the set of directed links composed of two ordered pairs of nodes, which represent the resource outflow from r, being an inflow to node ℓ, with $r, \ell \in \tilde{\mathcal{V}}$. On the other hand, consider a bidirectional communication network among the sub-systems, which determines the possible information sharing in order to compute the appropriate control inputs. The communication network is represented by an undirected graph denoted by $\mathcal{G} = (\mathcal{V}, \mathcal{E}, \mathbf{A})$, where $\mathcal{E} \subset \{(r, \ell) : r, \ell \in \mathcal{V}\}$ is the set of links allowing bidirectional communication between $r, \ell \in \mathcal{V}$, i.e., (r, ℓ) and (ℓ, r) represent the same link, and $\mathbf{A} \in \mathbb{R}^{n \times n}$ is the adjacency matrix, i.e., $a_{r\ell} = 1$ indicates that r and ℓ can share information. These two graphs $\tilde{\mathcal{G}}$ and $\mathcal{G}$ are presented in Fig. 7.1. Notice that storage elements in the distribution flow-based network presented in Fig. 7.1 may be considered as source elements supplying other lower-level storage elements as, in turn, are presented in Fig. 7.2.

Storage nodes have discharge coefficients given by $K_\ell > 0$, for all $\ell \in \mathcal{V}$, which determine their outflow resource. Furthermore, the inflows of the storage nodes $p_\ell \in \mathbb{R}_{>0}$, for all $\ell \in \mathcal{V}$, are manipulated imposing a proportion of resource, i.e., Qp_ℓ, where $0 < p_\ell \le 1$, for all $\ell \in \mathcal{V}$, and let $Q \in \mathbb{R}_{>0}$ be the total resource in the system. Besides, storage nodes have associated a vector of system states denoted by $\mathbf{z} \in \mathbb{R}_{>0}^n$ determining the amount of resource at each of these nodes. The storage nodes have the following dynamics:

$$\dot{z}_\ell = Qp_\ell - K_\ell z_\ell, \quad \forall \ell \in \mathcal{V}, \tag{7.1}$$

Fig. 7.3 Closed loop for the
sub-system corresponding to
the jth partition of the ith
topology

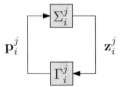

where the equilibrium duple $(p_\ell^\star, z_\ell^\star)$ implies that a non-null steady state has been
achieved for the stored resource, i.e., $z_\ell^\star > 0$ since $p_\ell^\star > 0$. In the aforementioned
flow-based network, the resource Q is distributed throughout the storage nodes,
which can be seen as sub-systems within the network. For this distribution system,
there are two objectives. First, it is desired to make an evenhanded distribution of a
resource throughout different sub-systems, i.e.,

$$\underset{p}{\text{minimize}} \left(z_\ell - \frac{1}{n} \sum_{\ell=1}^n z_\ell \right), \quad \forall \ell \in \mathcal{V},$$

while the second objective consists in determining the appropriate distribution of
costs for the sub-systems in function of their contribution to the first control objective, which is attained by using the available communication channels to coordinate
the distribution of the resource. The communication cost in a time interval $[t_0, t_f]$
associated to the use of $\mathcal{G} = (\mathcal{V}, \mathcal{E}, \mathbf{A})$ is given by

$$\text{KPI}_{\text{links}} = \frac{1}{2} \int_{t_0}^{t_f} \mathbb{1}_n^\top \mathbf{A} \mathbb{1}_n \ dt, \tag{7.2}$$

where t_0 and t_f denote the initial and final simulation time, respectively. In order
to achieve the second aforementioned objective, the *fair* cost $\tilde{C}_\ell$ that the ℓth sub-
system should pay for using the communication network must be found. It should
be satisfied that $\sum_{\ell \in \mathcal{V}} \tilde{C}_\ell = \text{KPI}_{\text{links}}$.

Finally, assume that the communication graph can have T different topologies
from the set $\mathcal{T} = \{1, \ldots, T\}$. Moreover, each topology $i \in \mathcal{T}$ has P_i different
partitions from the set $\mathcal{P}_i = \{1, \ldots, P_i\}$, where each partition $j \in \mathcal{P}_i$ of the topology
$i \in \mathcal{T}$ is a complete undirected graph denoted by $\mathcal{G}_i^j = (\mathcal{V}_i^j, \mathcal{E}_i^j)$, where $\mathcal{V}_i^j$ is the set
of n_i^j storage nodes within the corresponding partition, and $\mathcal{E}_i^j = \{(r, \ell) : r, \ell \in \mathcal{V}_i^j\}$
is the set of links representing the full information within each partition. Besides,
each partition $j \in \mathcal{P}_i$ of the topology $i \in \mathcal{T}$ represents a sub-system denoted by Σ_i^j,
which has a disjoint controller denoted by Γ_i^j as presented in Fig. 7.3.

The two different control objectives are achieved by using a game-theoretical
approach in a distributed manner. The first control objective associated to an evenhanded distribution of the resource is reached with a non-cooperative approach (using
a power index). In contrast, the determination of the *fair* distribution of costs for all
the sub-systems is made with a cooperative-game approach. Furthermore, suppose

that the topology of the communication graph can be reconfigured conveniently to achieve the control objectives every time $\tau > 0$ (this fact is discussed in Sect. 7.2).

A feature of the flow-based distribution network is its passivity property. Lemma 7.1 presents this property as in [8, 9].

Lemma 7.1 *The storage model of the flow-based distribution network Σ_i^j in Fig. 7.3 is passive defining its inputs as the error $e_{z_\ell} = z_\ell - z_\ell^\star$, and its outputs as the error $e_{p_\ell} = p_\ell - p_\ell^\star$, for all $\ell \in \mathcal{V}_i^j$, and for all partitions $j \in \mathcal{P}_i$ of the topology $i \in \mathcal{T}$.*

Proof The storage model dynamics in error coordinates, i.e., the error dynamics $\dot{e}_{z_\ell}$, are as follows:

$$\dot{e}_{z_\ell} = Q\left(e_{p_\ell} + p_\ell^\star\right) - K_\ell\left(e_{z_\ell} + z_\ell^\star\right), \quad \forall \ell \in \mathcal{V}_i^j.$$

Then, consider the storage function (also Lyapunov function)

$$E_1 = \frac{1}{2Q} \sum_{i \in \mathcal{T}} \sum_{j \in \mathcal{P}_i} \sum_{\ell \in \mathcal{V}_i^j} e_{z_\ell}^2, \tag{7.3}$$

whose derivative is given by

$$\dot{E}_1 = \frac{1}{Q} \sum_{i \in \mathcal{T}} \sum_{j \in \mathcal{P}_i} \sum_{\ell \in \mathcal{V}_i^j} e_{z_\ell} \dot{e}_{z_\ell},$$

$$= \frac{1}{Q} \sum_{i \in \mathcal{T}} \sum_{j \in \mathcal{P}_i} \sum_{\ell \in \mathcal{V}_i^j} e_{z_\ell} Q\left(e_{p_\ell} + p_\ell^\star\right) - e_{z_\ell} K_\ell\left(e_{z_\ell} + z_\ell^\star\right),$$

$$= \frac{1}{Q} \sum_{i \in \mathcal{T}} \sum_{j \in \mathcal{P}_i} \sum_{\ell \in \mathcal{V}_i^j} e_{z_\ell} Q e_{p_\ell} + e_{z_\ell} Q p_\ell^\star - e_{z_\ell}^2 K_\ell - e_{z_\ell} K_\ell z_\ell^\star,$$

$$= \frac{1}{Q} \sum_{i \in \mathcal{T}} \sum_{j \in \mathcal{P}_i} \sum_{\ell \in \mathcal{V}_i^j} e_{z_\ell} Q e_{p_\ell} - e_{z_\ell}^2 K_\ell + e_{z_\ell} \underbrace{\left(Q p_\ell^\star - K_\ell z_\ell^\star\right)}_{\dot{z}_\ell^\star = 0},$$

$$\leq \sum_{i \in \mathcal{T}} \sum_{j \in \mathcal{P}_i} \sum_{\ell \in \mathcal{V}_i^j} e_{z_\ell} e_{p_\ell},$$

$$= \sum_{i \in \mathcal{T}} \sum_{j \in \mathcal{P}_i} \sum_{\ell \in \mathcal{V}_i^j} \left(z_\ell - z_\ell^\star\right)\left(p_\ell - p_\ell^\star\right),$$

$$= \sum_{i \in \mathcal{T}} \sum_{j \in \mathcal{P}_i} \left(\mathbf{z}_i^j - \mathbf{z}_i^{j\star}\right)^\top \left(\mathbf{p}_i^j - \mathbf{p}_i^{j\star}\right).$$

Then, since $\left(\mathbf{z}_i^j - \mathbf{z}_i^{j\star}\right)$ are the inputs of $\boldsymbol{\Sigma}_i^j$ and $\left(\mathbf{p}_i^j - \mathbf{p}_i^{j\star}\right)$ are the outputs of $\boldsymbol{\Sigma}_i^j$,

then $\dot{E}_1 \leq \sum_{i \in \mathcal{T}} \sum_{j \in \mathcal{P}_i} \left(\mathbf{z}_i^j - \mathbf{z}_i^{j\star}\right)^\top \left(\mathbf{p}_i^j - \mathbf{p}_i^{j\star}\right)$ allows to conclude that storage-node

dynamics in the flow-based distribution network are passive.   □

7.2   Population Games in a New Context

Consider a population composed of a finite and large number of rational agents that make the decisions to select among a set of possible strategies $\mathcal{V} = \{1, \ldots, n\}$ (each strategy associated to a storage node, see Sect. 7.1). Agents change the strategy to improve their utilities or benefits.

Within the population there are $T \in \mathbb{Z}_{>0}$ possibly different topologies represented by a graph. This graph also determines how the population structure is configured, i.e., how the agents can interact among them. Each topology $i \in \mathcal{T}$ is given by a non–complete graph denoted by $\mathcal{G}_i = (\mathcal{V}, \mathcal{E}_i, \mathbf{A}_i)$, where $\mathcal{V}$ is the set of n nodes representing the strategies, the set of links that represent the possible interaction among agents selecting the corresponding strategies is denoted by $\mathcal{E}_i$, and $\mathbf{A}_i$ is the adjacency matrix of the corresponding ith topology.

Each topology $i \in \mathcal{T}$ has $P_i \in \mathbb{Z}_{>0}$ disjoint partitions. The set of partitions of the population topology $i \in \mathcal{T}$ is given by $\mathcal{P}_i = \{1, \ldots, P_i\}$. The partition $j \in \mathcal{P}_i$ of the topology $i \in \mathcal{T}$ is denoted by a complete graph $\mathcal{G}_i^j = (\mathcal{V}_i^j, \mathcal{E}_i^j)$, where $\mathcal{V}_i^j$ is the set of $n_i^j < n$ nodes representing the set of strategies within the corresponding partition $\mathcal{V}_i^j \subset \mathcal{V}$, and $\mathcal{E}_i^j$ is the set of $n_i^j(n_i^j - 1)/2$ links representing the full-information sharing and interaction within each partition. Furthermore, it must be satisfied that all the partitions form the entire topology, i.e., $\bigcup_{j \in \mathcal{P}_i} \mathcal{G}_i^j = \mathcal{G}_i$, for all $i \in \mathcal{T}$.

In the population, the scalar $p_\ell \in \mathbb{R}_{\geq 0}$ is the proportion of agents selecting the strategy $\ell \in \mathcal{V}$. The vector $\mathbf{p} \in \mathbb{R}_{\geq 0}^n$ is a population state or a strategic distribution composed of all the proportion of agents selecting the available strategies. The set of all the possible population states is given by a simplex denoted by $\Delta = \left\{\mathbf{p} \in \mathbb{R}_{\geq 0}^n : \mathbf{p}^\top \mathbb{1}_n = \pi\right\}$ with $\pi = 1$ for the application presented in this chapter. Similarly, $p_{i,\ell}^j \in \mathbb{R}_{\geq 0}$ is the proportion of agents selecting the strategy $\ell \in \mathcal{V}_i^j$ available in the partition $j \in \mathcal{P}_i$ of the topology $i \in \mathcal{T}$. The vector $\mathbf{p}_i^j \in \mathbb{R}_{\geq 0}^{n_i^j}$ is the strategic distribution of agents in the partition $j \in \mathcal{P}_i$ of the topology $i \in \mathcal{T}$. Finally, let π_i^j be the total mass in the partition $j \in \mathcal{P}_i$ of the topology $i \in \mathcal{T}$ given by $\pi_i^j = \mathbf{p}_i^{j\top} \mathbb{1}_{n_i^j}$. Since partitions are disjoint, then $\sum_{j \in \mathcal{P}_i} \pi_i^j = \pi$, with $\pi = 1$.

The payoff that agents receive for selecting a particular strategy is given by a fitness function $f_\ell : \Delta \to \mathbb{R}$, for the associated strategy $\ell \in \mathcal{V}$. The vector of fitness functions in the population, denoted by $\mathbf{f}$, is composed of all the fitness functions $f_\ell(\mathbf{p})$, $\ell \in \mathcal{V}$. Similarly, the vector of fitness functions of the partition $j \in \mathcal{P}_i$ of the topology $i \in \mathcal{T}$ is denoted by $\mathbf{f}_i^j$, which is composed of all the fitness functions

$f_\ell(\mathbf{p})$, $\ell \in \mathcal{V}_i^j$. The average fitness function in the population is given by $\bar{f} = \mathbf{p}^\top \mathbf{f}$. The average fitness for each partition of each topology is given by $\bar{f}_i^j = \left(\mathbf{p}_i^{j^\top} \mathbf{f}_i^j \right) / \pi_i^j$, for $j \in \mathcal{P}_i$, $i \in \mathcal{T}$, and the average fitness vector for the partition $j \in \mathcal{P}_i$ of the topology $i \in \mathcal{T}$ is $\bar{\mathbf{f}}_i^j = \mathbb{1}_{n_i^j} \bar{f}_i^j$.

In this chapter, both replicator and projection dynamics are used [10]. These two population dynamics are especially appealing since they have gradient properties discussed in [6].

Replicator and Projection Dynamics

For a fixed topology, there is a replicator dynamics system for each partition (see replicator dynamics in (2.14)). For a topology $i \in \mathcal{T}$ and a partition $j \in \mathcal{P}_i$, the replicator dynamics introduced in [11] are given by

$$\dot{\mathbf{p}}_i^j = \mathrm{diag}\left(\mathbf{p}_i^j \right) \left(\mathbf{f}_i^j - \bar{\mathbf{f}}_i^j \right), \tag{7.4}$$

and the projection dynamics introduced in [12] (see projection dynamics in (2.16)) are given by

$$\dot{\mathbf{p}}_i^j = \mathbf{f}_i^j - \frac{1}{n_i^j} \mathbb{1}_{n_i^j} \mathbf{f}_i^{j^\top} \mathbb{1}_{n_i^j}. \tag{7.5}$$

Then, the system changes among topologies[2] in order to use, at each iteration, a limited number of communication links. The equilibrium of interest in (7.4) for this chapter is the non-pure Nash equilibrium given by the condition $\mathbf{f}_i^{j\star} = \bar{\mathbf{f}}_i^{j\star}$, for all $j \in \mathcal{P}_i, i \in \mathcal{T}$. This is equivalent to $\mathbf{f}_i^{j\star} \in \mathrm{span}\{\mathbb{1}_{n_i^j}\}$, for all $i \in \mathcal{T}$ and $j \in \mathcal{P}_i$. Notice that the dynamics (7.4) have an equilibrium point $\mathbf{p}_i^{j\star} = 0$, which implies the extinction of the agents. Consequently, it is assumed that, under the replicator dynamics (7.4), there is not extinction of proportion of agents. Regarding the equilibrium for (7.5), it is achieved when all the fitness functions get the same value, i.e., $\mathbf{f}_i^{j\star} \in \mathrm{span}\{\mathbb{1}_{n_i^j}\}$, for all $i \in \mathcal{T}$ and $j \in \mathcal{P}_i$. Due to the fact that each topology is a non-connected graph, this equilibrium is achieved at each partition. Moreover, since topologies and partitions are varying over time, it is necessary to identify the equilibrium for all topologies and for all partitions, i.e., $f_k(\mathbf{p}^\star) = f_\ell(\mathbf{p}^\star)$, for all $k, \ell \in \mathcal{V}$.

Remark 7.1 Suppose that $\mathbf{f}$ is a stable game, then $(\mathbf{p} - \mathbf{p}^\star)^\top (\mathbf{f}(\mathbf{p}) - \mathbf{f}(\mathbf{p}^\star)) \leq 0$ according to Definition 2.2. Moreover, due to the fact that $\mathbf{f}(\mathbf{p}^\star) \in \mathrm{span}\{\mathbb{1}_n\}$, and that $\mathbf{p}, \mathbf{p}^\star \in \Delta$, then $(\mathbf{p} - \mathbf{p}^\star)^\top \mathbf{f}(\mathbf{p}^\star) = 0$. This leads to $(\mathbf{p} - \mathbf{p}^\star)^\top \mathbf{f}(\mathbf{p}) \leq 0$. ◇

The first property of both replicator and projection dynamics is the invariance of the simplex set Δ, which is analyzed in Proposition 7.1.

Proposition 7.1 *Let $\mathbf{p}(0)$ be the initial condition of (7.4) or (7.5). If $\mathbf{p}(0) \in \Delta$, then $\mathbf{p} \in \Delta$, for all $t \geq 0$, i.e., the simplex Δ is an invariant set under replicator dynamics (7.4) or projection dynamics (7.5), for any partition topology.*

---

[2]These topologies are determined by a partitioning performed by using a cooperative game approach, which is later discussed in this chapter.

Proof The invariant-set property is analyzed for both population dynamics.

Replicator dynamics: in order to prove that the simplex Δ is an invariant set, it is shown that the sum of $\dot{\mathbf{p}}_i^j$ for all topologies $i \in \mathcal{T}$ and for all partitions $j \in \mathcal{P}_i$ is null, i.e.,

$$
\sum_{i \in \mathcal{T}} \sum_{j \in \mathcal{P}_i} \mathbb{1}_{n_i^j}^\top \dot{\mathbf{p}}_i^j = \sum_{i \in \mathcal{T}} \sum_{j \in \mathcal{P}_i} \mathbf{p}_i^{j\top} \left(\mathbf{f}_i^j - \frac{1}{\pi_i^j} \mathbb{1}_{n_i^j} \mathbf{p}_i^{j\top} \mathbf{f}_i^j \right),
$$

$$
= \sum_{i \in \mathcal{T}} \sum_{j \in \mathcal{P}_i} \left(\mathbf{p}_i^{j\top} \mathbf{f}_i^j - \frac{1}{\pi_i^j} \mathbf{p}_i^{j\top} \mathbb{1}_{n_i^j} \mathbf{p}_i^{j\top} \mathbf{f}_i^j \right),
$$

$$
= \sum_{i \in \mathcal{T}} \sum_{j \in \mathcal{P}_i} \left(\mathbf{p}_i^{j\top} \mathbf{f}_i^j - \mathbf{p}_i^{j\top} \mathbf{f}_i^j \right),
$$

$$
= 0.
$$

Projection dynamics: in order to prove that the simplex Δ is an invariant set, it is shown that the sum of $\dot{\mathbf{p}}_i^j$ for all topologies $i \in \mathcal{T}$ and for all partitions $j \in \mathcal{P}_i$ is null, i.e.,

$$
\sum_{i \in \mathcal{T}} \sum_{j \in \mathcal{P}_i} \mathbb{1}_{n_i^j}^\top \dot{\mathbf{p}}_i^j = \sum_{i \in \mathcal{T}} \sum_{j \in \mathcal{P}_i} \mathbb{1}_{n_i^j}^\top \left(\mathbf{f}_i^j - \frac{1}{n_i^j} \mathbb{1}_{n_i^j} \mathbf{f}_i^{j\top} \mathbb{1}_{n_i^j} \right),
$$

$$
= \sum_{i \in \mathcal{T}} \sum_{j \in \mathcal{P}_i} \mathbb{1}_{n_i^j}^\top \mathbf{f}_i^j - \frac{1}{n_i^j} \mathbb{1}_{n_i^j}^\top \mathbb{1}_{n_i^j} \mathbf{f}_i^{j\top} \mathbb{1}_{n_i^j}.
$$

Notice that $\frac{1}{n_i^j} \mathbb{1}_{n_i^j}^\top \mathbb{1}_{n_i^j} = 1$, then

$$
\sum_{i \in \mathcal{T}} \sum_{j \in \mathcal{P}_i} \mathbb{1}_{n_i^j}^\top \dot{\mathbf{p}}_i^j = \sum_{i \in \mathcal{T}} \sum_{j \in \mathcal{P}_i} \mathbb{1}_{n_i^j}^\top \mathbf{f}_i^j - \mathbf{f}_i^{j\top} \mathbb{1}_{n_i^j},
$$

$$
= 0.
$$

Both results complete the proof.   □

Now, the passivity property of both the replicator and the projection dynamics is analyzed in Lemma 7.2.

Lemma 7.2 $\mathbf{\Gamma}_i^j$ *in Fig. 7.3 can be given by either the replicator dynamics (7.4) or the projection dynamics (7.5). For any of these dynamics, $\mathbf{\Gamma}_i^j$ is lossless defining its inputs as $\mathbf{f}_i^j$, and its outputs as $\mathbf{p}_i^j - \mathbf{p}_i^{j*}$, for all partitions $j \in \mathcal{P}_i$ of the topology $i \in \mathcal{T}$.*

Proof The analysis is made for both replicator and projection dynamics using different storage functions.

Replicator dynamics: consider the following entropy function as a storage function (also Lyapunov function), i.e.,

$$
\begin{aligned}
E_2(\mathbf{p}) &= -\sum_{i \in \mathcal{T}} \sum_{j \in \mathcal{P}_i} \sum_{\ell \in \mathcal{V}_i^j} p_{i,\ell}^{j\star} \ln \left(\frac{p_{i,\ell}^j}{p_{i,\ell}^{j\star}} \right), \\
&= -\sum_{i \in \mathcal{T}} \sum_{\ell \in \mathcal{V}} p_\ell^\star \ln \left(\frac{p_\ell}{p_\ell^\star} \right),
\end{aligned}
\tag{7.6}
$$

which is a valid Lyapunov function since $E_2(\mathbf{p}^\star) = 0$ and $E_2(\mathbf{p}) > 0$, for all $\mathbf{p} \neq \mathbf{p}^\star$, fact that is verified by using the Jensen's inequality (i.e., $G(g(x)) \geq g(G(x))$ for any convex function as the logarithm [13]), e.g., $E(\mathbf{p}) \geq -\sum_{i \in \mathcal{T}} \ln \left(\sum_{\ell \in \mathcal{V}} p_\ell^\star \frac{p_\ell}{p_\ell^\star} \right)$, and due to the fact that $\ln \left(\sum_{\ell \in \mathcal{V}} p_\ell^\star \frac{p_\ell}{p_\ell^\star} \right) = \ln(1)$, it follows that $E(\mathbf{p}) \geq 0$. Then $E(\mathbf{p}) > 0$, for all $\mathbf{p} \neq \mathbf{p}^\star$. Hence,

$$
\dot{E}_2(\mathbf{p}) = -\sum_{i \in \mathcal{T}} \sum_{j \in \mathcal{P}_i} \sum_{\ell \in \mathcal{V}_i^j} \frac{p_{i,\ell}^{j\star}}{p_{i,\ell}^j} \dot{p}_{i,\ell}^j.
\tag{7.7}
$$

Replacing dynamics (7.4) in (7.7) yields

$$
\begin{aligned}
\dot{E}_2(\mathbf{p}) &= -\sum_{i \in \mathcal{T}} \sum_{j \in \mathcal{P}_i} \mathbf{p}_i^{j\star\top} \left(\mathbf{f}_i^j - \frac{1}{\pi_i^j} \mathbb{1}_{n_i^j} \mathbf{p}_i^{j\top} \mathbf{f}_i^j \right), \\
&= -\sum_{i \in \mathcal{T}} \sum_{j \in \mathcal{P}_i} \left(\mathbf{p}_i^{j\star\top} - \frac{1}{\pi_i^j} \mathbf{p}_i^{j\star\top} \mathbb{1}_{n_i^j} \mathbf{p}_i^{j\top} \right) \mathbf{f}_i^j, \\
&= \sum_{i \in \mathcal{T}} \sum_{j \in \mathcal{P}_i} \left(\mathbf{p}_i^j - \mathbf{p}_i^{j\star} \right)^\top \mathbf{f}_i^j.
\end{aligned}
\tag{7.8}
$$

Projection dynamics: consider the quadratic error as a storage function (also Lyapunov function), i.e.,

$$
E_3(\mathbf{p}) = \frac{1}{2} \sum_{i \in \mathcal{T}} \sum_{j \in \mathcal{P}_i} \sum_{\ell \in \mathcal{V}_i^j} \left(p_{i,\ell}^j - p_{i,\ell}^{j\star} \right)^2,
\tag{7.9}
$$

where $E_3(\mathbf{p}) > 0$ for all $\mathbf{p} \neq \mathbf{p}^\star$, and $E_3(\mathbf{p}^\star) = 0$. Replacing dynamics (7.5) in the storage function derivative yields

$$\dot{E}_3(\mathbf{p}) = \sum_{i \in \mathcal{T}} \sum_{j \in \mathcal{P}_i} \left(\mathbf{p}_i^j - \mathbf{p}_i^{j\star}\right)^\top \left(\mathbf{f}_i^j - \frac{1}{n_i^j} \mathbb{1}_{n_i^j} \mathbf{f}_i^{j\top} \mathbb{1}_{n_i^j}\right),$$

$$= \sum_{i \in \mathcal{T}} \sum_{j \in \mathcal{P}_i} \left(\mathbf{p}_i^j - \mathbf{p}_i^{j\star}\right)^\top \mathbf{f}_i^j - \frac{1}{n_i^j} \left(\mathbf{p}_i^j - \mathbf{p}_i^{j\star}\right)^\top \mathbb{1}_{n_i^j} \mathbf{f}_i^{j\top} \mathbb{1}_{n_i^j},$$

$$= \sum_{i \in \mathcal{T}} \sum_{j \in \mathcal{P}_i} \left(\mathbf{p}_i^j - \mathbf{p}_i^{j\star}\right)^\top \mathbf{f}_i^j - \frac{1}{n_i^j} \left(\mathbf{p}_i^{j\top} \mathbb{1}_{n_i^j} - \mathbf{p}_i^{j\star\top} \mathbb{1}_{n_i^j}\right) \mathbf{f}_i^{j\top} \mathbb{1}_{n_i^j},$$

$$= \sum_{i \in \mathcal{T}} \sum_{j \in \mathcal{P}_i} \left(\mathbf{p}_i^j - \mathbf{p}_i^{j\star}\right)^\top \mathbf{f}_i^j.$$

Then, since $\mathbf{f}_i^j$ are the inputs and $\left(\mathbf{p}_i^j - \mathbf{p}_i^{j\star}\right)$ are the outputs, both results show that the replicator and projection dynamics are lossless. $\qquad\square$

Remark 7.2 According to Remark 7.1 and taking into account the derivative of the storage functions (also Lyapunov functions) $\dot{E}_2 \leq 0$, and $\dot{E}_3 \leq 0$, there exists a time $\tau > 0$ such that $\|p_\ell(t) - p_\ell^\star(t)\| \geq \|p_\ell(t + \tau) - p_\ell^\star(t + \tau)\|$, for all $\ell \in \mathcal{V}$. $\qquad\diamond$

Once the passivity features of the considered flow-based distribution network and both the replicator and projection dynamics have been presented, it is shown that the equilibrium point of the closed-loop system in Fig. 7.3 is stable as stated in Proposition 7.2.

Proposition 7.2 *The equilibrium pair* $(\mathbf{z}_i^{j\star}, \mathbf{p}_i^{j\star})$, *for all partitions* $j \in \mathcal{P}_i$ *in the topology* $i \in \mathcal{T}$ *(i.e., the equilibrium* $\mathbf{z}_i^{j\star}$ *of the sub-system* Σ_i^j, *and the equilibrium* $\mathbf{p}_i^{j\star}$ *of the system* Γ_i^j, *for the closed loop presented in Fig. 7.3) is stable under the replicator and projection dynamics selecting* $\mathbf{f}_i^j = \mathbf{z}_{\max,i}^j - \mathbf{z}_i^j$, *where* $\mathbf{z}_{\max,i}^j$ *is the constant maximum capacity of the storage nodes belonging to* $\mathcal{V}_i^j$.

Proof Consider the Lyapunov function given by $E = E_1 + E_2$ for the replicator dynamics, and $E = E_1 + E_3$ for the projection dynamics, where E_1, E_2 and E_3 are the functions presented in (7.3), (7.6), and (7.9), respectively. According to Lemmas 7.1 and 7.2, for both cases $\dot{E}$ satisfies

$$\dot{E} \leq \sum_{i \in \mathcal{T}} \sum_{j \in \mathcal{P}_i} \left(\mathbf{z}_i^j - \mathbf{z}_i^{j\star}\right)^\top \left(\mathbf{p}_i^j - \mathbf{p}_i^{j\star}\right) + \sum_{i \in \mathcal{T}} \sum_{j \in \mathcal{P}_i} \left(\mathbf{p}_i^j - \mathbf{p}_i^{j\star}\right)^\top \mathbf{f}_i^j. \qquad (7.10)$$

Adding and substracting $\sum_{i \in \mathcal{T}} \sum_{j \in \mathcal{P}_i} \mathbf{z}_{\max,i}^j$ in (7.10) yields

$$\dot{E} \leq \sum_{i \in \mathcal{T}} \sum_{j \in \mathcal{P}_i} \left(-\mathbf{z}_{\max,i}^j + \mathbf{z}_i^j + \mathbf{z}_{\max,i}^j - \mathbf{z}_i^{j\star} \right)^\top \left(\mathbf{p}_i^j - \mathbf{p}_i^{j\star} \right) + \sum_{i \in \mathcal{T}} \sum_{j \in \mathcal{P}_i} \left(\mathbf{p}_i^j - \mathbf{p}_i^{j\star} \right)^\top \mathbf{f}_i^j,$$

$$= \sum_{i \in \mathcal{T}} \sum_{j \in \mathcal{P}_i} \left\{ \left(-\mathbf{z}_{\max,i}^j + \mathbf{z}_i^j \right)^\top \left(\mathbf{p}_i^j - \mathbf{p}_i^{j\star} \right) + \left(\mathbf{z}_{\max,i}^j - \mathbf{z}_i^{j\star} \right)^\top \left(\mathbf{p}_i^j - \mathbf{p}_i^{j\star} \right) + \mathbf{f}_i^{j\top} \left(\mathbf{p}_i^j - \mathbf{p}_i^{j\star} \right) \right\},$$

$$= \sum_{i \in \mathcal{T}} \sum_{j \in \mathcal{P}_i} \left\{ -\mathbf{f}_i^{j\top} \left(\mathbf{p}_i^j - \mathbf{p}_i^{j\star} \right) + \mathbf{f}_i^{j\star\top} \left(\mathbf{p}_i^j - \mathbf{p}_i^{j\star} \right) + \mathbf{f}_i^{j\top} \left(\mathbf{p}_i^j - \mathbf{p}_i^{j\star} \right) \right\},$$

$$= 0.$$

Therefore, the equilibrium point $(\mathbf{z}_i^{j\star}, \mathbf{p}_i^{j\star})$, for all partitions $j \in \mathcal{P}_i$ in the topology $i \in \mathcal{T}$, is stable. $\qquad\square$

Notice that the population-games-based controller in Fig. 7.3 is a data-driven controller since it is designed without requiring the model of the system.

7.3 Coalitional-Game Role and Partitioning Criterion

Consider a cooperative game with transferable utility defined as a pair $(\mathcal{V}, V_c)$, where $\mathcal{V} = \{1, .., n\}$ is the set of players (each player associated to a storage node, see Sect. 7.1), and V_c is the characteristic function. From the cooperative-game viewpoint for each topology $i \in \mathcal{T}$, each node $\ell \in \mathcal{V}$ is a player and each partition $j \in \mathcal{P}_i$ represents a coalition of players.

7.3.1 Computation of the Shapley Value

The characteristic function V_c assigns a real value to each of the $2^\mathcal{V}$ coalitions and returns a real value. Formally, the characteristic function is a mapping $V_c : 2^\mathcal{V} \to \mathbb{R}$. For each coalition, $\mathcal{O} \subseteq \mathcal{V}$, $V_c(\mathcal{O})$ is the value that players can share among themselves. Additionally, for the empty coalition, $V_c(\emptyset) \triangleq 0$.

Prior to defining the characteristic function, costs associated to each coalition are defined as

$$\tilde{C}(\mathcal{O}) = \frac{1}{|\mathcal{O}|} \sum_{\ell \in \mathcal{O}} \tilde{C}_\ell, \tag{7.11}$$

where $\tilde{C}_\ell$ is the individual cost of player belonging to the coalition $\ell \in \mathcal{O}$. For the considered flow-based distribution networks, costs are defined to be $\tilde{C}_\ell = z_{\max,\ell} - z_\ell$, for all $\ell \in \mathcal{V}$, where $z_{\max,\ell}$ is the maximum capacity of the storage node, and z_ℓ is the current state of the corresponding storage node. This individual value represents a cost that a player would have to assume in case that it does not cooperate with anyone. Notice that the error $\tilde{C}_\ell$ is an appropriate selection for the costs since a player ℓ with null error does not have incentives to cooperate with others due to the fact it has achieved the first control objective, and cooperation would imply to increment the error. In contrast, a player with a big error $\tilde{C}_\ell$ has incentives to cooperate in order to minimize its error and should assume costs associated to that cooperation. In addition, if two players have identical errors $\tilde{C}_\ell$, then it is reasonable that both assume the same costs to cooperate each other. In other words, it is reasonable that those players with bigger errors incur in higher costs to achieve the first control objective.

Furthermore, the individual cost allows the player to determine its incentives to establish a coalition with another player, i.e., the player would have incentives to cooperate with others as long as the cooperation implies a reduction of its costs. Besides, these individual costs allow computing the contribution of a player to a coalition, i.e., how the cost is reduced as the player collaborates with an existing coalition. Afterwards, the specific considered characteristic function is given by the difference between the sum of individual costs and the cost of the corresponding coalition. Then, the difference between the costs when each player operates by itself and the costs when all these players collaborate with each other can be used to define the cost function as

$$V_c(\mathcal{O}) = \sum_{\ell \in \mathcal{O}} \tilde{C}_\ell - \tilde{C}(\mathcal{O}). \tag{7.12}$$

A solution of the cooperative game is an allocation rule that provides each player with a payoff according to its contribution. Let $\mathbf{y} \in \mathbb{R}^n$ be the payoff vector given by $\mathbf{y} = [y_1 \quad \cdots \quad y_n]^\top$. Some desirable properties for the distribution of the $V_c(\mathcal{O})$ among the players are:

1. Efficiency: $\sum_{\ell \in \mathcal{O}} y_\ell \le V_c(\mathcal{V})$,
2. Coalitional rationality: $\sum_{\ell \in \mathcal{O}} y_\ell = V_c(\mathcal{O})$ for all coalitions $\mathcal{O} \subseteq \mathcal{V}$,
3. Individual rationality: $y_\ell \ge V_c(\{\ell\})$, for all $\ell \in \mathcal{V}$.

A payoff rule that satisfies the mentioned desired requirements is the Shapley value (or Shapley power index) [14, 15], which is given by

$$\Phi(\ell) = \sum_{\mathcal{O} \subseteq \mathcal{V} \setminus \{\ell\}} \Psi(\mathcal{O})\big(V_c(\mathcal{O} \cup \{\ell\}) - V_c(\mathcal{O})\big), \tag{7.13}$$

where

$$\Psi(\mathcal{O}) = \frac{|\mathcal{O}|!\,(n - |\mathcal{O}| - 1)!}{n!}.$$

Notice that the sum in (7.13) considers all the possible coalitions where player $\ell \in \mathcal{V}$ can be added. Its computation requires full information from all the players and coalitions of the cooperative game, resulting in high computational burden. In particular, the combinatorial explosion when having a high number of players is a common issue in this context.

Once the characteristic function has been defined as in (7.12), a mathematical relationship between the Shapley values can be determined to mitigate the high computational cost enhancing the possibilities to use a distributed structure. This result is presented in Theorem 7.1.

Theorem 7.1 *Let (7.12) be the characteristic function of a cooperative game with the set of players $\mathcal{V} = \{1, ..., n\}$. Let $\tilde{C}_\ell$ be the cost associated to each player $\ell \in \mathcal{V}$, and (7.11) be the costs associated to each coalition $\mathcal{O} \subseteq \mathcal{V}$. The Shapley value $\Phi(\ell)$ for any player $\ell \in \mathcal{V}$ is computed as follows:*

$$\Phi(\ell) = \frac{1}{n}\left(V_c(\mathcal{V}) - \Theta\left(\sum_{r \in \mathcal{V}\setminus\{\ell\}} \tilde{C}_r - (n-1)\tilde{C}_\ell\right)\right), \qquad (7.14)$$

where $\Theta > 0$ is a constant for the cooperative game whose value only depends on the number of players n, i.e.,

$$\Theta = \sum_{s=1}^{n-2}\left\{\left(\frac{(n-2)!}{s!(n-2-s)!}\right)\left(\frac{s}{s+1}\right)\left(\frac{s!\,(n-s-1)! + (s+1)!\,(n-s-2)!}{n!}\right)\right\}.$$

Besides, there is a relationship between Shapley values given by $\Phi(r) = \Phi(\ell) + (\tilde{C}_r - \tilde{C}_\ell)\Theta$, for all $r, \ell \in \mathcal{V}$.

Proof First, it is proven the relationship between the Shapley values of different players with the constant Θ given by $\Phi(r) = \Phi(\ell) + (\tilde{C}_r - \tilde{C}_\ell)\Theta$, for all $r, \ell \in \mathcal{V}$. The Shapley value $\Phi(\ell)$ of the player $\ell \in \mathcal{V}$ in (7.13) may be re-written by expressing the set of coalitions to which the player $\ell \in \mathcal{V}$ can be added, in terms of a second player $r \in \mathcal{V}$, as follows:

$$\Phi(\ell) = \sum_{\mathcal{O} \subseteq \mathcal{V}\setminus\{\ell,r\}} \Psi(\mathcal{O})\,(V_c(\mathcal{O} \cup \{\ell\}) - V_c(\mathcal{O})) +$$

$$\sum_{\mathcal{O} \subseteq \mathcal{V}\setminus\{\ell,r\}} \Psi(\mathcal{O} \cup \{r\})\,(V_c(\mathcal{O} \cup \{r\} \cup \{\ell\}) - V_c(\mathcal{O} \cup \{r\})).$$

Similarly, the Shapley value $\Phi(r)$ of the player $r \in \mathcal{V}$ may be written in terms of player $\ell \in \mathcal{V}$ as follows:

$$\Phi(r) = \sum_{\mathcal{O} \subseteq \mathcal{V}\setminus\{\ell,r\}} \Psi(\mathcal{O})\,(V_c(\mathcal{O} \cup \{r\}) - V_c(\mathcal{O})) +$$

$$\sum_{\mathcal{O} \subseteq \mathcal{V}\setminus\{\ell,r\}} \Psi(\mathcal{O} \cup \{\ell\})\,(V_c(\mathcal{O} \cup \{r\} \cup \{\ell\}) - V_c(\mathcal{O} \cup \{\ell\})).$$

Now, it is found the difference between the Shapley values $\Phi(r)$ and $\Phi(\ell)$, denoted by $\tilde{\Phi}(r, \ell) = \Phi(r) - \Phi(\ell)$. Here, it is taken into account that $\Psi(\mathcal{O} \cup \{r\}) = \Psi(\mathcal{O} \cup \{\ell\})$. Hence

$$\tilde{\Phi}(r, \ell) = \sum_{\mathcal{O} \subseteq \mathcal{V} \backslash \{\ell, r\}} \Psi(\mathcal{O}) \{V_c(\mathcal{O} \cup \{r\}) - V_c(\mathcal{O} \cup \{\ell\})\} +$$

$$\sum_{\mathcal{O} \subseteq \mathcal{V} \backslash \{\ell, r\}} \Psi(\mathcal{O} \cup \{r\}) \{V_c(\mathcal{O} \cup \{r\}) - V_c(\mathcal{O} \cup \{\ell\})\}.$$

$$(7.15)$$

Replacing (7.12) and (7.11) in (7.15), it is obtained

$$\tilde{\Phi}(r, \ell) = \sum_{\mathcal{O} \subseteq \mathcal{V} \backslash \{\ell, r\}} \Psi(\mathcal{O}) \left\{ \left(1 - \frac{1}{|\mathcal{O}| + 1}\right) (\tilde{C}_r - \tilde{C}_\ell) \right\} +$$

$$\sum_{\mathcal{O} \subseteq \mathcal{V} \backslash \{\ell, r\}} \Psi(\mathcal{O} \cup \{r\}) \left\{ \left(1 - \frac{1}{|\mathcal{O}| + 1}\right) (\tilde{C}_r - \tilde{C}_\ell) \right\}.$$

Briefly, the difference between the Shapley values $\tilde{\Phi}(r, \ell) = \Phi(r) - \Phi(\ell)$ is given by

$$\tilde{\Phi}(r, \ell) = (\tilde{C}_r - \tilde{C}_\ell) \underbrace{\sum_{\mathcal{O} \subseteq \mathcal{V} \backslash \{\ell, r\}} \overbrace{(\Psi(\mathcal{O}) + \Psi(\mathcal{O} \cup \{r\}))}^{\theta_1} \overbrace{\left(\frac{|\mathcal{O}|}{|\mathcal{O}| + 1}\right)}^{\theta_2}}_{\Theta}.$$

Notice that the constant value Θ can be re-written as

$$\Theta = \sum_{s=1}^{n-2} \left\{ \overbrace{\left(\frac{(n-2)!}{s!(n-2-s)!}\right)}^{\theta_3} \overbrace{\left(\frac{s}{s+1}\right)}^{\theta_2} \underbrace{\left(\frac{s!(n-s-1)! + (s+1)!(n-s-2)!}{n!}\right)}_{\theta_1} \right\},$$

where θ_3 represents the amount of coalitions that can be formed in the cooperative game with s players, i.e., $|\mathcal{O}| = s$. Finally,

$$\Phi(r) = \Phi(\ell) + (\tilde{C}_r - \tilde{C}_\ell)\Theta, \quad \forall r \in \mathcal{V} \backslash \{\ell\}.$$

It follows that making the sum over all the elements of the set $\mathcal{V} \backslash \{\ell\}$, it is obtained the desired relationship. i.e.,

$$\sum_{r \in \mathcal{V} \backslash \{\ell\}} \Phi(r) = (n-1)\Phi(\ell) + \Theta \sum_{r \in \mathcal{V} \backslash \{\ell\}} \tilde{C}_r - \Theta(n-1)\tilde{C}_\ell.$$

Then, if $\Phi(\ell)$ is added at both sides of the previous equality, it yields

$$\sum_{r \in \mathcal{V}} \Phi(r) = n\Phi(\ell) + \Theta\left(\sum_{r \in \mathcal{V}} \tilde{C}_r - (n-1)\tilde{C}_\ell\right),$$

and since $V_c(\mathcal{V}) = \sum_{r \in \mathcal{V}} \Phi(r)$, then

$$\Phi(\ell) = \frac{1}{n}\left(V_c(\mathcal{V}) - \Theta\left(\sum_{r \in \mathcal{V}\setminus\{\ell\}} \tilde{C}_r - (n-1)\tilde{C}_\ell\right)\right),$$

completing the proof. □

Remark 7.3 If $\tilde{C}_r > \tilde{C}_\ell$, then $\Phi(r) > \Phi(\ell)$, for all $r, \ell \in \mathcal{V}$. Suppose that $\tilde{C}_r > \tilde{C}_\ell$, and since $\Theta > 0$, then $\Phi(r) = \Phi(\ell) + (\tilde{C}_r - \tilde{C}_\ell)\Theta > \Phi(\ell)$. ◇

7.3.2 Computation of the Shapley Value Under a Distributed Structure

The reduction of the computational time according to the result obtained in Theorem 7.1 allows investing extra computational effort to gather all the needed information to compute the Shapley value under a distributed structure. In order to make the distributed Shapley computation, let one player be a *pivot* player in charge of collecting all the required information from all the other players known as *supply* players.

Remark 7.4 Even though the computation of the Shapley value is performed by one player, the required structure is a non-complete graph. In this regard, the computation of the Shapley value is made under a distributed structure. ◇

Consider a connected non-complete graph for the distributed computation of the Shapley value denoted by $\mathcal{G} = (\mathcal{V}, \mathcal{E})$, where $\mathcal{V}$ is the set of nodes representing the players within the cooperative game. Moreover, let $q \in \mathcal{V}$ denote the pivot player, and $\mathcal{V}\setminus\{q\}$ is the set of supply players. The set of links $\mathcal{E}$ represents possible communication among the players. Moreover, let $\mathcal{N}_\ell$ be the set of neighbors of player $\ell \in \mathcal{V}$, i.e., $\mathcal{N}_\ell = \{r : (\ell, r) \in \mathcal{E}\}$. Figure 7.4 shows the case for a path graph where the first player is the pivot player, i.e., $q = 1 \in \mathcal{V}$. Notice that the required information to compute the Shapley value is given by all the individual costs of all the supply players according to (7.11) (since the pivot player knows its own individual cost),

pivot

Fig. 7.4 Path graph for the distributed Shapley computation. Without loss of generality, the pivot player is considered to be player 1

and the total number of players. This information is required by the pivot player who is in charge of the Shapley value computation in order to perform the partitioning (which is used for establishing the proper topologies introduced in Sect. 7.2) and also assign the fair distribution of costs associated to the communication links. It is necessary that the pivot player obtains the information in a distributed way subject to the communication topology given by the graph $\mathcal{G}$.

The distributed algorithm is inspired by the work presented in [16], and it is split into two different tasks, i.e., the distributed computation of the number of players n, and the distributed computation of all the individual costs for all the supply players $\tilde{C}_\ell$, for all $\ell \in \mathcal{V}\backslash\{q\}$ as it has been presented in Chap. 4. Each stage of the algorithm is presented next.

7.3.2.1   Computation of the Number of Players

Consider an auxiliary variable for the pivot player ξ_q, and for each supply player ξ_ℓ, where $\ell \in \mathcal{V}\backslash\{q\}$. The initial conditions of these auxiliary variables are given by $\xi_q(0) = 1$, and $\xi_\ell(0) = 0$, for all $\ell \in \mathcal{V}\backslash\{q\}$. The following consensus algorithm can be implemented taking advantage of the relationship between the initial conditions and the stationary point as [17, 18]

$$\dot{\xi}_\ell = \sum_{r \in \mathcal{N}_\ell} (\xi_r - \xi_\ell), \quad \forall \ell \in \mathcal{V}. \tag{7.16}$$

Since the communication graph $\mathcal{G}$ among players is connected, according to [17] the stationary value of (7.16) is $\xi_\ell^\star = (\sum_{r \in \mathcal{V}} \xi_r(0))/n$, for all $\ell \in \mathcal{V}$. Consequently, the stationary value is given by $\xi_\ell^\star = n^{-1}$, for all $\ell \in \mathcal{V}$. This fact shows that the pivot player can get information about the total number of players n in a distributed way.

7.3.2.2   Computation of the Individual Costs

For simplicity, it is assumed that the pivot player is the player $q = 1 \in \mathcal{V}$ (see, e.g., Fig. 7.4). Since there are $n - 1$ values for individual costs that should be sent to the pivot player, then there is a turn value denoted by $\kappa \in \mathbb{Z}_{>0}$. The token assigns a flag for each player to distribute its individual cost. For the case with a pivot player $q = 1 \in \mathcal{V}$, the turn value κ should vary from 2 to n in order to cover the total number of players in the cooperative game. After determining the possible values for the turn variable κ, i.e., $\kappa = 2, \ldots, n$, this variable is initialized for the first supply player, i.e., $\kappa = 2$. Once the pivot player determines the individual cost corresponding to this supply player, the turn value is increased, i.e., $\kappa = \kappa + 1$ in order to allow the next supply player distribute its individual cost. This process is repeated until the last player distributes its cost to the pivot player, i.e., until $\kappa = n$.

Consider the auxiliary variables ψ_ℓ, for all $\ell \in \mathcal{V}$. The initial conditions of these auxiliary variables are given by $\psi_\kappa = \tilde{C}_\kappa$ from (7.11), and $\psi_\ell = 0$, for all $\ell \in \mathcal{V}\backslash\{\kappa\}$.

Then, the consensus algorithm [17] is implemented, i.e,

$$\dot{\psi}_\ell = \sum_{r \in \mathcal{N}_\ell} (\psi_r - \psi_\ell), \quad \forall \ell \in \mathcal{V}. \tag{7.17}$$

The stationary value is given by $\psi_\ell^\star = \frac{1}{n} \sum_{r \in \mathcal{V}} \psi_r(0)$. For the selected initial conditions, $\psi_\ell^\star = n^{-1} \tilde{C}_\kappa$. Furthermore, since the number of players is already known from the procedure presented in Sect. 7.3.2.1, then

$$\psi_\ell^\star = \tilde{C}_\kappa \xi_\ell^\star, \quad \forall \kappa \in \mathcal{V} \backslash \{q\}.$$

It is concluded that, finding the number of players and finding each individual cost by assigning turns to the supply players, it is possible to compute the Shapley value in a distributed way.

Remark 7.5 Notice that it is necessary to compute n distributed consensus algorithms, one algorithm for the determination of the number of players, and $n - 1$ algorithms to distribute the individual costs of supply players to the pivot player. This fact implies an extra computational burden. However, it is shown that the distributed computation of the Shapley value using distributed consensus and Theorem 7.1 is lower than the computational burden of the classical approach (7.13) as the total number of players is increased.   ◇

In order to verify the difference between the computational burden of computing the Shapley value with (7.13), and by using the relation proposed in Theorem 7.1 with the constant value Θ, different Shapley values for several amount of players have been computed.

The comparison among the computation of the Shapley value in a distributed and centralized way using Theorem 7.1, and in a centralized way using (7.13), is presented in Table 7.1. Moreover, Fig. 7.5 shows the summary of the computational burden for all the approaches and for different number of players n.

7.3.3 Partitioning Procedure

In order to perform the partitioning of the system, first it is established a time τ that satisfies Remark 7.2, which defines when the proper topology is evaluated.[3] Every time τ, the Shapley value $\Phi(\ell)$ of all the players $\ell \in \mathcal{V}$ is computed by using the low-computational-cost operation with the factor Θ. A second necessary parameter is the size of the desired partitions, denoted by $\tilde{g} \in \mathbb{Z}_{>0}$, i.e., make partitions with a number $\tilde{g} < n$ of players.

---

[3]The proper topology is determined by considering a fair distribution of costs given by the Shapley value, which establishes the appropriate partitioning within the system at every time τ that satisfies Remark 7.2.

Table 7.1 Comparison of computational burden for computing the Shapley value for different number of players

Total Number of players n	Centralized approach by using (7.13) time [s]	Centralized approach by using (7.14) time [ms]	Distributed approach by using (7.14) time [s]
3	0.4232	0.09	2.7899
4	0.8020	0.12	3.6002
6	1.2907	0.18	5.4838
8	2.3421	0.24	7.2707
10	5.9998	0.30	9.0004
12	25.6436	0.36	10.8073
14	110.6065	0.42	12.6004
16	1944.7919	0.48	14.4005
18	53938.9433	0.54	16.2210
100	–	3	90.0030

Fig. 7.5 Computational burden for computing the Shapley value (semi-log plot). Centralized computation with typical approach using (7.13), centralized computation with the proposed approach using (7.14), and with distributed computation with the proposed approach using (7.14)

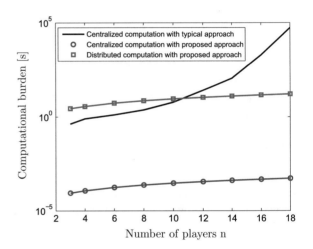

In the partitioning procedure, it is desired to gather the player with the highest Shapley value with the $\tilde{g} - 1$ players with the lowest Shapley values. In this sense, it is possible to make a cooperation in which the best players share their benefits with those in a worst situation. Details of this partitioning process at every time τ are presented in Algorithm 3. It is worth to highlight that this partitioning criterion might be different depending on the control objectives and the system dynamical behavior. In this chapter, the partitioning is performed in function of the Shapley value by grouping players with the highest power index with those with lowest power index. This procedure allows players to unify their power index and therefore to achieve an evenhanded distribution of resource.

Algorithm 3 Partitioning algorithm based on the Shapley value

---

1: **procedure** INITIALIZATION
2:    $\tilde{g} \leftarrow$ desired number of players per partition
3:    $\{\mathcal{H}\}_\ell \leftarrow \Phi(\ell)$ set of the Shapley values $|\mathcal{H}| = n$
4:    $j \leftarrow 1$ index for partitions
5: **end procedure**
6: **while** $|\mathcal{H}| \geq \tilde{g}$ **do**
7:    $b \leftarrow 0$ flag for amount of players
8:    $\mathcal{C}_j \leftarrow \emptyset$ initialization of the jth partition
9:    $r \leftarrow \arg\max_{\ell \in \mathcal{V}} \{\mathcal{H}\}_\ell$ player with maximum value
10:    $\mathcal{H} \leftarrow \mathcal{H} \backslash \{\mathcal{H}\}_r$ reduce the set without rth value
11:    $\mathcal{C}_j \leftarrow \mathcal{C}_j \cup \{r\}$
12:    $b \leftarrow b + 1$ number of players in $\mathcal{C}_j$
13:    **while** $b \leq \tilde{g}$ **do**
14:       $r \leftarrow \arg\min_{\ell \in \mathcal{V}} \{\mathcal{H}\}_\ell$ player with minimum value
15:       $\mathcal{H} \leftarrow \mathcal{H} \backslash \{\mathcal{H}\}_r$ reduce the set without rth value
16:       $\mathcal{C}_j \leftarrow \mathcal{C}_j \cup \{r\}$
17:       $b \leftarrow b + 1$ number of players in $\mathcal{C}_j$
18:    **end while**
19:    $j \leftarrow j + 1$ number of partitions
20: **end while**
21: **if** $\mathcal{H} \neq \emptyset$ **then**
22:    $\mathcal{C}_j \leftarrow \mathcal{H}$ leftover players forming a smaller partition
23: **end if**

Lemma 7.3 *The stable closed-loop system presented in Fig. 7.3, changing the partitions every τ by using the Shapley values, converges to the common equilibrium $z_\ell^\star = z_r^\star$ for all $r, \ell \in \mathcal{V}$.*

Proof Consider the arbitrary initial condition for the storage nodes $\mathbf{z}(0) \in \mathbb{R}_{\geq 0}^n$, the maximum capacities for the storage nodes $\mathbf{z}_{\max} \in \mathbb{R}_{\geq 0}^n$, and the costs $\tilde{C}_i(0) = z_{\max,i} - z_i(0)$, for all $i \in \mathcal{V}$. Now, consider $r^0 = \arg\max_{i \in \mathcal{V}} \tilde{C}_i(0)$, and $\ell^0 = \arg\min_{i \in \mathcal{V}} \tilde{C}_i(0)$. Then, it is known that according to Remark 7.3, $\Phi(r^0) > \Phi(\ell^0)$ and that $r^0, \ell^0 \in \mathcal{V}$ are assigned to the same partition according to the proposed partitioning procedure (Algorithm 3). It follows that, during the interval time τ with the closed loop presented in Fig. 7.3 and according to Remark 7.2, it is obtained $\tilde{C}_{\ell^0} \leq z_{\max,r^0} - z_{r^0}(\tau) \leq \tilde{C}_{r^0}$, and $\tilde{C}_{\ell^0} \leq z_{\max,r^0} - z_{r^0}(\tau) \leq \tilde{C}_{r^0}$. Then computing $r^1 = \arg\max_{i \in \mathcal{V}} \tilde{C}_i(\tau)$, and $\ell^1 = \arg\min_{i \in \mathcal{V}} \tilde{C}_i(\tau)$, it is obtained that $\tilde{C}_{\ell^0} \leq \tilde{C}_{\ell^1}$, and $\tilde{C}_{r^1} \leq \tilde{C}_{r^0}$.
Making the partitioning, it follows that $\tilde{C}_{\ell^0} \leq \tilde{C}_{\ell^1} \leq z_{\max,\ell^1} - z_{\ell^1}(2\tau) \leq \tilde{C}_{r^1} \leq \tilde{C}_{r^0}$, and $\tilde{C}_{\ell^0} \leq \tilde{C}_{\ell^1} \leq z_{\max,r^1} - z_{r^1}(2\tau) \leq \tilde{C}_{r^1} \leq \tilde{C}_{r^0}$. It is concluded that

$$\min_{i \in \mathcal{V}} \tilde{C}_i(k\tau) \leq z_{\max,\ell} - z_\ell((k+1)\tau) \leq \max_{i \in \mathcal{V}} \tilde{C}_i(k\tau),$$

$$\min_{i \in \mathcal{V}} \tilde{C}_i(k\tau) \leq z_{\max,r} - z_r((k+1)\tau) \leq \max_{i \in \mathcal{V}} \tilde{C}_i(k\tau),$$

where $r = \arg\max_{i \in \mathcal{V}} \tilde{C}_i(k\tau)$, and $\ell = \arg\min_{i \in \mathcal{V}} \tilde{C}_i(k\tau)$. Then $[\max_{i \in \mathcal{V}} \tilde{C}_i(k\tau)$ $- \min_{i \in \mathcal{V}} \tilde{C}_i(k\tau)] \to 0$ as $k \to \infty$, until $\max_{i \in \mathcal{V}} \tilde{C}_i(k\tau) = \min_{i \in \mathcal{V}} \tilde{C}_i(k\tau)$. This situation leads to the equilibrium $z_r^\star = z_\ell^\star$, for all $r, \ell \in \mathcal{V}$. Consequently, it is obtained that all fitness functions are the same, then x_ℓ^* is achieved for all $\ell \in \mathcal{V}$.   □

7.4   Case Study: Water Supply Network

As a case study, a multi-objective problem involving both competition and cooperation is presented. The case study is shown in Fig. 7.6, which is composed of $n = 13$
tanks. There are two control objectives. The former objective is to maintain all the
tanks at the same maximum level, which is solved by finding a Nash equilibrium.
The latter objective is to determine the cost each player should pay for these communication links according to its contribution to achieve the desired equilibrium. The
fair cost that each player should pay is given by using the Shapley value.

Let $\mathrm{KPI}_{\mathrm{links}}$ in (7.2) denote the communication cost associated to the undirected
graph $\mathcal{G}$ (see Fig. 7.1b). Therefore, the fair cost distribution is given by the Shapley
value, i.e.,

$$\tilde{L}_\ell = \frac{\Phi(\ell)}{\sum_{r=1}^n \Phi(r)} \mathrm{KPI}_{\mathrm{links}}, \quad \ell \in \mathcal{V}, \tag{7.18}$$

where $\tilde{L}_\ell$ is the fair cost corresponding to the player $\ell \in \mathcal{V}$. The Shapley value is normalized in (7.18) since it is computed in terms of the error rather than in economical
units. These communication links allow sharing, in a local way, information about
the measured levels in order to achieve the first control objective. Notice that this is
a decentralized control scheme since each control input is computed by using only
partial and local information.

In order to solve the control problem, it is proposed to make a partitioning based on
the Shapley value as presented in Sect. 7.3. Therefore, the individual cost associated
to each player is the error between the current level and the maximum level of each
tank, i.e.,

$$\tilde{C}_\ell = h_{\max,\ell} - h_\ell, \tag{7.19}$$

where $h_{\max,\ell}$ denotes the maximum and constant possible level for the ℓth tank, and
h_ℓ is the current measured level of the ℓth tank. Once the partitions are determined, a
population-dynamics approach is applied to each partition, where the fitness function

Fig. 7.6 Case study:
Resource Q allocation
throughout n tanks

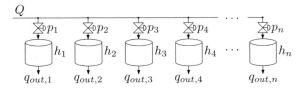

for each strategy is given by the error defined according to Proposition 7.2, which is equal to the cost $\bar{C}_\ell$, i.e., $f_\ell = h_{\max,\ell} - h_\ell$, for all $\ell \in \mathcal{V}$. It is assumed that this fitness function is always non-negative since physically the current measured level is $h_\ell \leq h_{\max,\ell}$. Notice that, with this fitness function, more resource is assigned to those tanks with lower level. The dynamics for the ℓth tank are given by

$$\dot{h}_\ell = q_{in,\ell} - q_{out,\ell},$$
$$q_{out,\ell} = K_\ell h_\ell,$$

where h_ℓ is the current level, $q_{in,\ell}$ is the inflow, $q_{out,\ell}$ is the outflow, and K_ℓ is a constant factor characterizing the outflow (discharge coefficient), for the ℓth tank. There is a constant available resource given by $Q = 30\,\mathrm{m}^3/\mathrm{s}$. Each inflow $q_{in,\ell}$ is controlled by a valve commanded by a control signal p_ℓ, i.e., $q_{in,\ell} = Qp_\ell$, with $0 \leq p_\ell \leq 1$. It is assumed that there is a local controller at each valve that guarantees the desired flow given by p_ℓ. The limited resource establishes a constraint over all the inflows, i.e., $\sum_{\ell=1}^{n} q_{in,\ell} = Q$. This equality constraint leads to the condition $\sum_{\ell=1}^{n} Qp_\ell = Q$, and consequently, $\sum_{\ell=1}^{n} p_\ell = 1$, then $\mathbf{p} \in \Delta$. This condition is also satisfied in the partitioned system if the initial condition of the proportion of agents satisfies $\sum_{j \in \mathcal{P}_i} \mathbb{1}_{n_j}^\top \mathbf{p}_i^j(0) = 1$, for the initial topology $i \in \mathcal{T}$, due to Proposition 7.1.

In order to analyze the performance of the closed-loop system composed of the flow-based distribution network and the population dynamics, three KPIs are stated.

- The costs associated to the communication links φ_{links} is taken as a KPI given by

$$\text{KPI}_{\text{links}} = \frac{1}{2} \int_{t_0}^{t_f} \mathbb{1}_n^\top \mathbf{A}_i \mathbb{1}_n \, dt, \quad i \in \mathcal{T}, \tag{7.20}$$

which is the same KPI presented in (7.2) but for the graph $\mathcal{G}_i$ corresponding to the current topology $i \in \mathcal{T}$, where t_0 is an initial time and t_f a final time.
- The error of each tank level with respect to the average levels of the system tanks is also considered as a performance indicator, i.e.,

$$\text{KPI}_{\text{error},\ell} = \int_{t_0}^{t_f} \left(h_\ell - \frac{1}{n} \mathbf{h}^\top \mathbb{1}_n \right) dt, \quad \ell \in \mathcal{V},$$

where $\mathbf{h} = [h_1 \quad \dots \quad h_n]^\top$. Then, there is a KPI for the whole system in function of the mentioned error levels, i.e.,

$$\text{KPI}_{\text{error}} = \sum_{\ell=1}^{n} \text{KPI}_{\text{error},\ell}.$$

- The settling time of the system states (level of tanks) with criterion of the 5%, i.e.,

$$\text{KPI}_{\text{settling}} = \min_{t} \{ t : \tilde{t} \geq t, \ \underline{\mathbf{h}} \leq \mathbf{h}(\tilde{t}) \leq \bar{\mathbf{h}} \},$$

Fig. 7.7 Complete graph
given by the grand coalition

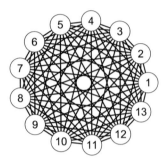

where $\underline{\mathbf{h}} = 0.95\mathbf{h}^{\star}$, $\bar{\mathbf{h}} = 1.05\mathbf{h}^{\star}$, with $\mathbf{h}^{\star}$ corresponds to the equilibrium point of the system.

7.4.1 Results

As a reference to analyze the performance of the proposed control strategy, results for the centralized case are also presented. This is equivalent to the case with a topology given by a complete graph (see Fig. 7.7).

For this example, $\tau = 2.5$ s, and the Shapley value is computed as a function of the error costs in (7.19). It is important to mention that the initial condition for each tank level has been established as a random value from the interval [0, 1]. The objectives are:

(i) maintain all the tank levels at the same maximum value; and
(ii) determine the fair costs for the players.

Figure 7.8 shows the evolution of the tank levels for three different cases (partitions of 2 players, partitions of 3 players and the grand partition), where it is also shown

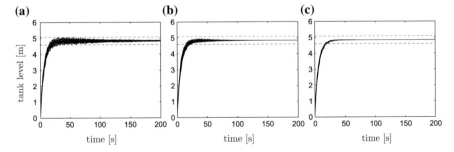

Fig. 7.8 Evolution of the 13 tank levels for three different cases: **a** partitions of two players, **b** partitions of three players, and **c** partition of n players (full information). Initial condition for each tank level has been determined by a random value within the interval [0, 1]

Table 7.2 Fair economical costs for each player determined with the Shapley value

Player $\ell \in \mathcal{V}$	Fair costs according Shapley value $(\Phi(\ell)\text{KPI}_{\text{links}})/(\sum_{r=1}^{n} \Phi(r))$	
	Partitions 2 Players	Partitions 3 Players
1	101.9982	162.7683
2	103.4772	164.7666
3	103.0618	175.4056
4	67.3064	161.5046
5	69.9416	294.2980
6	69.7096	146.2820
7	78.9709	215.1184
8	123.0623	167.3544
9	91.6612	290.9280
10	76.7914	146.8018
11	99.1411	154.4804
12	75.6172	170.1767
13	139.2561	150.1145
Total	1200	2400

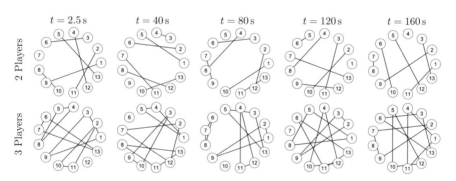

Fig. 7.9 Evolution of the graph topologies given by partitions of two and three players (i.e., $\tilde{g} = 2$, and $\tilde{g} = 3$), for five different iterations

the gap that determines the settling time. Every τ, the topology of the system is determined. In Fig. 7.8, it can be also seen that the first objective is met for all the cases. The second objective is achieved by determining the Shapley value as presented in Table 7.2. Figure 7.9 shows some different topologies obtained based on the Shapley value (see Algorithm 3) at five different time instants. In that figure, it can be seen how topologies vary dynamically with partitions of two and three players.

Figure 7.10 shows a sensitivity analysis for the evolution of the tank levels for $n_s = 10$ different scenarios. The only difference among these ten scenarios is the vector of initial conditions, which are selected randomly within the interval $[0, 2]$. First, it is presented the mean of all the tank levels for the scenarios, i.e.,

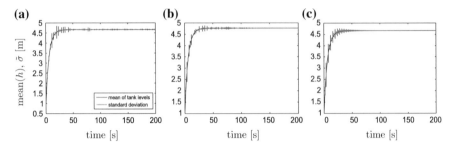

Fig. 7.10 Sensitivity analysis made based on scenarios with random initial conditions with $n = 13$ tank levels. Mean evolution of the tank levels and their corresponding standard deviation along the time for three different cases: **a** partitions of two players, **b** partition of three players, and **c** partition of n players (full information)

$$\tilde{\varrho}(h(w\tilde{\delta})) = \frac{1}{n_s \, n} \sum_{s=1}^{n_s} \sum_{i=1}^{n} h_i^s(w\tilde{\delta}),$$

where $w = 1, \ldots, 2000$, are the data used to make the sensitivity analysis, $\tilde{\delta} = 0.1$ s in order to cover the whole simulation time (i.e., $t_f = 200$ s), and h_i^s corresponds to the tank $i \in \mathcal{V}$ in the sth simulation. Then, the standard deviation along the time is presented to analyze the transitory event for each one of the three cases, partitions of two players, three players, and the grand partition. The standard deviation is computed as follows:

$$\tilde{\sigma}(w\tilde{\delta}) = \sqrt{\frac{1}{n_s \, n} \sum_{s=1}^{n_s} \sum_{i=1}^{n} \left(h_i^s(w\tilde{\delta}) - \tilde{\varrho}(h(w\tilde{\delta})) \right)^2}.$$

It can be seen in Fig. 7.10 that the standard deviation is smaller as there is an evolution of the tank levels. Besides, the standard deviations are reduced more quickly when there are larger partitions. This fact is because when having bigger partitions, there are more exchange of communication among different strategies in the population game achieving the closed-loop system equilibrium in a faster way. The graph sequence presented in Fig. 7.9 varies as the initial conditions are modified because the Shapley value depends on them. However, the sensitivity analysis presented in Fig. 7.10 shows that the system achieves the equilibrium point for all the different initial conditions.

Finally, the comparison of the three cases with the proposed KPI is presented in Table 7.3. In these results, it can be seen the proper performance of the proposed methodology with respect to the centralized case with full information (partition of n players). Economical costs are reduced significantly, but it implies an increment of the transitory errors and settling time of the evolution of the system states (tank levels).

Table 7.3 Closed-loop system performance for different topologies. Coalitions of two and three players, and with full information

Players per partitions $\tilde{g}$	Communication KPI$_{\text{links}}$	Error KPI$_{\text{error}}$	Settling time KPI$_{\text{settling}}$
2 players	1200	197.1742	33.23
3 players	2400	58.1723	21.80
n players	15600	20.9294	19.73

7.4.2   Discussion

In general, game theory models the interaction between rational players. The two different main approaches, the cooperative- and the non-cooperative-game directions, solve quite different problems. One approach addresses problems in which there is competition whereas the other one is more appropriate to model cooperation. As it has been discussed here, it is possible to face engineering applications by implementing both approaches in different but related objectives.

The combination of game theoretical approaches (cooperative and non-cooperative) implies that the elements within the system (storage, sink and source elements) take different roles depending on the game theoretical perspective. For instance, in the presented flow-based distribution network of Sect. 7.1, storage nodes represent strategies in the non-cooperative-game approach. In contrast, storage nodes assume the role of players that demand information by using the communication network in the cooperative game. This multi-role situation is quite interesting due to the fact that each role is tightly related to each particular control objective, i.e., the optimal resource distribution for the non-cooperative game and both the fair distribution of costs and proper partitioning for the cooperative game.

It has been shown that the closed-loop system with the non-cooperative approach (i.e., with both the replicator and the projection dynamics) is stable, performing the partitioning criteria based on cooperative games. This is achieved based on the assumption that all the partitions, for all the topologies, form individually a complete graph. Additionally, when selecting conveniently the characteristic function of the cooperative game, it is possible to compute the Shapley value of the whole set of players from the Shapley value of any arbitrary player by following a polynomial-time procedure. Moreover, it has been shown that, by taking advantage of the coalition-rationality axiom, it is possible to compute the Shapley value by solving a set of linear equations. This result allows to reduce considerably the computational burden from an exponential-time procedure to a polynomial one. Besides, due to the fact that the computation can be reduced, it is also possible to spend time to compute the Shapley value under a distributed structure. For example, the comparison of the computational burden in Fig. 7.5 shows less computation cost for the distributed structure with respect to the traditional Shapley equation and a case involving more than ten players.

7.5 Summary

A control problem involving two objectives associated to competition and cooperation has been presented. In particular, a flow-based distribution network composed
of different sub-systems (storage nodes) has been considered. The first objective is
associated to an optimal distribution of the available resource, i.e., it is desired to
achieve a maximum and equal amount of resource throughout all the sub-systems
(storage nodes). To this end, a communication network allows the interaction among
different storage nodes, whose communication links have an economical cost. As a
second objective, it is desired to distribute the costs of the communication network
throughout the sub-systems. To this end, more influent tanks (with higher power
index) must pay less than those with lower influence (with lower power index). In
this regard, costs are assigned in a *fair* way for each storage node according to
their contributions to the achievement of the first objective. A non-cooperative-game
approach is used to solve the former objective, whereas a cooperative-game approach
is proposed to solve the latter objective. Furthermore, for the non-cooperative-game
approach, the stability of the closed-loop system is analyzed to guarantee that the first
objective is achieved. On the other hand, regarding the cooperative-game approach
that determines the *fair* distribution of communication costs, an alternative way to
compute the Shapley value for the selected specific characteristic function has been
presented in order to mitigate the computational burden. The proposed approach
also allows the computation of the Shapley value under distributed structures within
shorter computational time in comparison with the standard computation.

Moreover, bigger partitions imply more information exchange among the storage
nodes, and therefore the resource-distribution objective is achieved faster. Simulation results have shown that having more communication links imply shorter settling
time. When having large partitions, the dynamic resource allocation is made faster
due to the fact that there is more available information in the non-cooperative-game
approach. Nevertheless, when having more players making partitions, more communication links are required and this implies higher economical costs. Furthermore, a
smaller overshoot and settling time for the system states (tank levels) are obtained
when partitions are larger. Hence, there is a need for balancing economical cost and
performance in relation to the complexity of the system and the number of players
involved in the cooperative game to perform the partitioning. Finally, the sensitivity
analysis for the evolution of the system states with arbitrary initial conditions shows
that, despite of the fact that the partitioning highly depends on the initial conditions
of the system, the equilibrium of the closed-loop system is always achieved.

This chapter has used the concept of partitioning, which is obtained as a product of
the power index value, but it is not the main purpose of the game-theoretical approach.
Moreover, it is worth to highlight that the partitioning task and/or decomposition of
systems is crucial in the design of distributed controllers, and it should be determined
considering several aspects, e.g., dynamical coupling, coupled constraints, and/or
balance size among sub-system states. Therefore, next chapter is devoted to the
development of a methodology to perform partitioning of large-scale systems.

References

1. Owen G, Shapley LS (1989) Optimal location of candidates in ideological space. Int J Game Theory 18(3):339–356
2. Ramirez-Llanos E, Quijano N (2010) A population dynamics approach for the water distribution problem. Int J Control 83:1947–1964
3. Khalil HK (2002) Nonlinear systems. Prentice Hall, Upper Saddle River
4. Costa-Castelló R, Griñó R (2006) A repetitive controller for discrete-time passive systems. Automatica 42:1605–1610
5. Barreiro-Gomez J, Ocampo-Martinez C, Maestre JM, Quijano N (2015) Multi-objective model-free control based on population dynamics and cooperative games. In: Proceedings of the 54th IEEE conference on decision and control (CDC). Osaka, Japan, pp 5296–5301
6. Sandholm WH, Dokumaci E, Lahkar R (2008) The projection dynamic and the replicator dynamic. Games Econ Behav 64:666–683
7. Grosso J (2015) Economic and robust operation of generalised flow-based networks. Doctoral dissertation. Automatic control department, Universidad Politècnica de Catalunya
8. Barreiro-Gomez J, Obando G, Riaño-Briceño G, Quijano N, Ocampo-Martinez C (2015) Decentralized control for urban drainage systems via population dynamics: Bogota case study. In Proceedings of the European control conference (ECC). Linz, Austria, pp 2431–2436
9. Fox MJ, Shamma JS (2013) Population games, stable games, and passivity. Games 4(4):561–583
10. Sandholm WH (2010) Population games and evolutionary dynamics. MIT Press, Cambridge, Mass
11. Taylor PD, Jonker LB (1978) Evolutionary stable strategies and game dynamics. Math Biosci 40(1):145–156
12. Nagurney A, Zhang D (1997) Projected dynamical systems in the formulation, stability analysis, and computation of fixed demand traffic network equilibria. Transp Sci 31:147–158
13. Jensen JLWV (1906) Sur les fonctions convexes et les inégalités entre les valeurs moyennes. Acta Math 30(1):175–193
14. Owen G (1995) Game theory. Academic Press, Cambridge ISBN 9780125311519
15. Shapley LS (1953) A value for n-person games. Ann Math Stud 28:307–317
16. Garin F, Schenato L (2010) A survey on distributed estimation and control applications using linear consensus algorithms. Netw Control Syst 406:75–107
17. Olfati-Saber R, Fax JA, Murray RM (2007) Consensus and cooperation in networked multi-agent systems. Proc IEEE 95(1):215–233
18. Cai K, Ishii H (2014) Average consensus on arbitrary strongly connected digraphs with time-varying topologies. IEEE Trans Autom Control 59:1066–1071

Part III
Large-Scale Systems Partitioning in Control

Chapter 8
Partitioning for Large-Scale Systems: Sequential DMPC Design

Chapter 7, where the Shapley value has been studied, has also discussed a partitioning as a result of the power index [1]. On the contrary, this chapter is devoted to the development of a partitioning strategy. As it has been reviewed in Chap. 2, several partitioning methods have been discussed in literature for both particular and general cases, e.g., [2–9], among others. More precisely, the main contribution presented in this chapter is a novel partitioning approach of a non-directed graph representing information sharing inspired by the Kernighan-Lin algorithm [10], considering four different objectives, i.e., to minimize the number of links connecting partitions, to minimize the difference of the size of partitions, to minimize the distance among elements composing each partition, and to minimize the amount of relevant information that connects different partitions, i.e., it is also considered how relevant the information that a link provides is. Furthermore, prioritization weights assign importance to each objective as desired. Most of the partitioning methods consider a graph representation, i.e., algorithms consider graphs associated to the dynamics of the system. Differently, this chapter proposes to generate a graph that describes the information dependence among variables considered in the control design. As an application to illustrate the advantages of the partitioning approach addressed by using an information representation instead of a dynamical-model representation, a large-scale water supply system is considered, and an MPC controller is designed. To this end, the information graph is computed in order to determine an appropriate partitioning by using the proposed algorithm.

8.1 Large-Scale System Partitioning

Consider an undirected connected graph denoted by $\mathcal{G} = (\mathcal{S}, \mathcal{E})$ representing the topology of an information-sharing network, which is determined depending on the information that a control strategy requires to compute the appropriate control

© Springer International Publishing AG, part of Springer Nature 2019
J. Barreiro-Gomez, *The Role of Population Games in the Design of Optimization-Based Controllers*, Springer Theses,
https://doi.org/10.1007/978-3-319-92204-1_8

inputs, where $S = \{1, \ldots, n\}$ is the set of $n \in \mathbb{Z}_{>0}$ nodes representing variables considered within the control strategy, and $\mathcal{E} \subset \{(i, j) : i, j \in S\}$ is the set of links of the graph representing possible information sharing among nodes. The set of neighbors of a node $i \in S$ is given by $\mathcal{N}_i = \{j : (i, j) \in \mathcal{E}\}$. The graph is undirected since it is assumed that links represent bidirectional-information channels. In the information-sharing network, it is defined a distance function among nodes from the set S whose mapping is given by $\tilde{d} : S \times S \rightarrow \mathbb{R}_{\geq 0}$. Let $\mathbf{D} \in \mathbb{R}_{\geq 0}^{n \times n}$ be the constant matrix representing distances, i.e., $\tilde{d}_{ij} = \tilde{d}(i, j)$. Additionally, consider a function whose mapping is $c : S \times S \rightarrow \mathbb{R}_{\geq 0}$, where $c(i, j)$ determines how relevant the information shared between node $j \in S$ and node $i \in S$ is. Therefore, let $\mathbf{C} \in \mathbb{R}_{\geq 0}^{n \times n}$ be the constant matrix representing all the relevance factors, i.e., $c_{ij} = c(i, j)$. On the other hand, let $\mathcal{K} = \{1, \ldots, m\}$ be the set of indices for the $m \in \mathbb{Z}_{>0}$ partitions of the graph $\mathcal{G}$. The partitioning at time instant k is represented by the set $\mathcal{P}_k = \{S_k^\ell : \ell \in \mathcal{K}\}$, i.e., each partition of $\mathcal{G}$ at time instant k is an undirected connected graph of the form $\mathcal{G}_k^\ell = (S_k^\ell, \mathcal{E}_k^\ell)$, for all $\ell \in \mathcal{K}$, where $\bigcap_{\ell \in \mathcal{K}} S_k^\ell = \emptyset$, and $\bigcup_{\ell \in \mathcal{K}} S_k^\ell = S$, for all k. Given a partition $\mathcal{P}_k$, consider the function $g : S \rightarrow \mathcal{K}$ that receives a node $i \in S$ and returns the index $\ell \in \mathcal{K}$ that allows to identify to which partition the node $i \in S$ belongs to, i.e., $g(i) = \{\ell \in \mathcal{K} : i \in S_k^\ell\}$. Finally, let $\mathbf{V}_k \in \mathbb{R}^{n \times n}$ be the time-varying matrix whose element $v_{ij,k} = |S_k^{g(i)}| - |S_k^{g(j)}|$ if $g(i) \neq g(j)$, and nodes i, and j are neighbors, i.e., $j \in \mathcal{N}_i$.

Remark 8.1 Notice that if $\mathcal{P}_k$ is an admissible partition, i.e., $\bigcap_{\ell \in \mathcal{K}} S_k^\ell = \emptyset$, and $\bigcup_{\ell \in \mathcal{K}} S_k^\ell = S$, then the set $\{\ell \in \mathcal{K} : i \in S_k^\ell\}$ is singleton, for all $\ell \in \mathcal{K}$, and $g(i)$ is a function, for all $i \in S$.   ◇

8.1.1 Partitioning Problem Statement

The $m-$partitioning problem consists in finding the optimal set of partitions denoted by $\mathcal{P}^\star$ such that the following objectives are minimized:

Links[1]: minimize the amount of links connecting different partitions given by σ_1, i.e., links $(i, j) \in \mathcal{E}$ such that $(i, j) \notin \mathcal{E}_k^\ell$, for all $\ell \in \mathcal{K}$,

$$\sigma_1(\mathcal{P}_k) = \frac{1}{2} \sum_{\ell \in \mathcal{K}} \sum_{i \in S_k^\ell} \sum_{j \in S \setminus S_k^\ell} a_{ij}.$$

Nodes[2]: minimize the difference between the amount of nodes in the partitions $|S_k^\ell|$, for all $\ell \in \mathcal{K}$, and the average of total nodes of the graph $\mathcal{G}$ given by $\frac{n}{m}$, i.e., $\frac{1}{m} \sum_{\ell \in \mathcal{K}} ||S_k^\ell| - \frac{n}{m}|$. Notice that this objective may be expressed as the minimization of a function σ_2 depending on the time-varying matrix $\mathbf{V}_k$, i.e.,

---

[1]Presented as external balance in [2].

[2]Presented as internal balance in [2].

$$\sigma_2(\mathcal{P}_k) = \sum_{\ell \in \mathcal{K}} \sum_{i \in \mathcal{S}_k^\ell} \sum_{j \in \mathcal{S} \backslash \mathcal{S}_k^\ell} v_{ij,k}.$$

Distance: minimize the distance among the nodes belonging to the same partition, i.e., $\frac{1}{2} \sum_{\ell \in \mathcal{K}} \sum_{i \in \mathcal{S}_k^\ell} \sum_{j \in \mathcal{S}_k^\ell \backslash \{i\}} \tilde{d}_{ij}$. This objective may be expressed conveniently considering the inverse distance of links that are connecting different partitions, i.e.,

$$\sigma_3(\mathcal{P}_k) = \frac{1}{2} \sum_{\ell \in \mathcal{K}} \sum_{i \in \mathcal{S}_k^\ell} \sum_{j \in \mathcal{S} \backslash \mathcal{S}_k^\ell} \tilde{d}_{ij}^{-1}.$$

Relevance: minimize the information relevance of the links that are connecting different partitions, i.e.,

$$\sigma_4(\mathcal{P}_k) = \sum_{\ell \in \mathcal{K}} \sum_{i \in \mathcal{S}_k^\ell} \sum_{j \in \mathcal{S} \backslash \mathcal{S}_k^\ell} c_{ij}.$$

The four aforementioned objectives are prioritized by setting the vector of weights denoted by $\varphi \in \mathbb{R}_{\geq 0}^4$. The optimal partitioning $\mathcal{P}^\star$ is obtained by solving the following optimization problem:

$$\underset{\mathcal{P}}{\text{minimize}} \sum_{j=1}^4 \varphi_j \sigma_j(\mathcal{P}_k), \tag{8.1a}$$

$$\text{subject to} \bigcap_{\ell \in \mathcal{K}} \mathcal{S}_k^\ell = \emptyset, \tag{8.1b}$$

$$\bigcup_{\ell \in \mathcal{K}} \mathcal{S}_k^\ell = \mathcal{S}. \tag{8.1c}$$

8.1.2 Distributed Partitioning Algorithm

In order to solve the optimization problem (8.1), consider the weighted graph $\mathcal{G}$ where $\mathbf{W}_k \in \mathbb{R}^{n \times n}$ is a time-varying weighting matrix, i.e., $\mathcal{G} = (\mathcal{S}, \mathcal{E}, \mathbf{W}_k)$. The elements of $\mathbf{W}_k$ are denoted by $w_{ij,k} \in \mathbb{R}_{\geq 0}$ representing a cost associated to the link $(i, j) \in \mathcal{E}$, where $w_{ij,k} = \varphi_1 a_{ij} + \varphi_2 v_{ij,k} + \varphi_3 d_{ij}^{-1} + \varphi_4 c_{ij}$. Then, it is proposed to solve the $m-$partitioning problem (8.1) as follows:

$$\underset{\mathcal{P}}{\text{minimize}} \sum_{\ell \in \mathcal{K}} \sum_{i \in \mathcal{S}_k^\ell} \sum_{j \in \mathcal{S} \backslash \mathcal{S}_k^\ell} w_{ij,k}, \tag{8.2}$$

and subject to constraints (8.1b) and (8.1c). The set of nodes $\mathcal{S}_k^\ell$ of the subgraph $\mathcal{G}_k^\ell$ is composed of a set of internal nodes denoted by $\check{\mathcal{S}}_k^\ell$, and a set of external nodes denoted by $\hat{\mathcal{S}}_k^\ell$, for all $\ell \in \mathcal{K}$. The internal nodes from the set $\check{\mathcal{S}}_k^\ell$ only have connection to nodes that belong to the same partition. In contrast, the external nodes from the set $\hat{\mathcal{S}}_k^\ell$ have connection to at least one node that belongs to a different partition. Formally,

$$\check{\mathcal{S}}_k^\ell = \{i \in \mathcal{S}_k^\ell : \mathcal{N}_i \subseteq \mathcal{S}_k^\ell\}, \quad \forall \, \ell \in \mathcal{K},$$
$$\hat{\mathcal{S}}_k^\ell = \{i \in \mathcal{S}_k^\ell : \mathcal{N}_i \nsubseteq \mathcal{S}_k^\ell\}, \quad \forall \, \ell \in \mathcal{K},$$
$$\mathcal{S}_k^\ell = \check{\mathcal{S}}_k^\ell \cup \hat{\mathcal{S}}_k^\ell, \quad \forall \, \ell \in \mathcal{K}.$$

Each external node $i \in \hat{\mathcal{S}}_k^\ell$, for all $\ell \in \mathcal{K}$, represents a decision maker that is able to select a partition from the set

$$\mathcal{Q}_{i,k} = \{g(j) : j \in \mathcal{N}_i\} \backslash \{g(i)\}.$$

Moreover, each decision maker $i \in \{\cup_{\ell \in \mathcal{K}} \hat{\mathcal{S}}_k^\ell\}$, at time instant k, has associated an internal cost denoted by $\check{h}_i$, i.e.,

$$\check{h}_i(\mathcal{P}_k) = \sum_{j \in \mathcal{S}_k^{g(i)} \cap \mathcal{N}_i} w_{ij,k}, \quad \forall \, i \in \hat{\mathcal{S}}_k^\ell, \text{ and } \ell \in \mathcal{K},$$

and an external benefit denoted by $\hat{h}_i^\ell$, for each partition ℓ from the set $\mathcal{Q}_{i,k}$, i.e.,

$$\hat{h}_i^\ell(\mathcal{P}_k) = \sum_{j \in \mathcal{S}_k^\ell \cap \mathcal{N}_i} w_{ij,k}, \quad \forall \, i \in \hat{\mathcal{S}}_k^\ell, \text{ and } \ell \in \mathcal{Q}_{i,k}.$$

Then, the best external benefit for the decision maker $i \in \hat{\mathcal{S}}_k^\ell$, for all $\ell \in \mathcal{Q}_{i,k}$, is obtained as follows:

$$\hat{h}_i(\mathcal{P}_k) = \max_{\ell \in \mathcal{Q}_{i,k}} \hat{h}_i^\ell(\mathcal{P}_k).$$

Finally, as previously mentioned, the decision maker selects among the possible available partitions depending on a utility denoted by η_i, i.e., if the decision maker has incentives to move from one partition to another one, then

$$\eta_i(\mathcal{P}_k) = \max \left(0, \hat{h}_i(\mathcal{P}_k) - \check{h}_i(\mathcal{P}_k)\right), \quad \forall \, i \in \hat{\mathcal{S}}_k^\ell, \text{ and } \ell \in \mathcal{K}.$$

Further, consider an undirected graph $\tilde{\mathcal{G}}_k = (\tilde{\mathcal{S}}_k, \tilde{\mathcal{E}}_k)$ at time instant k –not necessarily connected– composed of all the decision makers –external nodes of $\mathcal{G}$–. Then, the set of nodes of $\tilde{\mathcal{G}}_k$ is given by $\tilde{\mathcal{S}}_k = \bigcup_{\ell \in \mathcal{K}} \hat{\mathcal{S}}_k^\ell$, and the set of links is given by $\tilde{\mathcal{E}}_k \subset \{(i, j) : i \in \hat{\mathcal{S}}_k^\ell, j \in \hat{\mathcal{S}}_k^r, \ell \neq r\}$. Let $\tilde{\mathbf{A}}_k \in \{0, 1\}^{|\tilde{\mathcal{S}}_k| \times |\tilde{\mathcal{S}}_k|}$ be the adjacency matrix of the graph. Since $\tilde{\mathcal{G}}_k$ is not necessarily a connected graph, it has $q \in \mathbb{Z}_{>0}$ components at time instant k, where the set of components of the graph is $\mathcal{C}_k = \{1, \ldots, q\}$. Each component is a graph denoted by $\tilde{\mathcal{G}}_k^z = (\tilde{\mathcal{S}}_k^z, \tilde{\mathcal{E}}_k^z)$, with adjacency matrix $\tilde{A}_k^z$, is connected and not necessarily complete, for all $z \in \mathcal{C}_k$. Furthermore, only the decision maker with higher incentives (winner decision maker in the component $z \in \mathcal{C}_k$ denoted by $i_k^{z\star} \in \mathcal{W}_k^z \subseteq \tilde{\mathcal{S}}_k^z$) would make a decision to switch from its current partition to another one among the set of available partitions, i.e.,

$$i_k^{z\star} \in \arg\max_{i \in \tilde{\mathcal{S}}_k^z} \; \eta_i(\mathcal{P}_k) = \mathcal{W}_k^z, \quad \forall \, z \in \mathcal{C}_k. \tag{8.3}$$

Remark 8.2 The decision maker $i_k^{z\star}$, for all $z \in \mathcal{C}_k$, can be computed as in (8.3) in a distributed manner satisfying the information-sharing graph $\tilde{\mathcal{G}}_k^z$, for all $z \in \mathcal{C}_k$. This procedure is made by using maximum consensus as in [11]. Therefore, the proposed partitioning algorithm can be performed in a distributed manner. ◇

Notice that (8.3) should be solved at each time instant $k \in \mathbb{Z}_{\geq 0}$. The best option for the decision maker $i_k^{z\star} \in \tilde{\mathcal{S}}_k^z$, for all $z \in \mathcal{C}_k$, to select a new partition is

$$\ell_k^{z\star} \in \underset{\ell \in \mathcal{Q}_{i_k^{z\star},k}}{\arg\max} \; \hat{h}_{i_k^{z\star}}^\ell(\mathcal{P}_k). \tag{8.4}$$

Hence, the partitioning is modified only if $\eta_i(\mathcal{P}_k) > \varkappa$, where $\varkappa \in \mathbb{R}_{>0}$ establishes a ending-up condition. The updating is as follows:

$$\mathcal{S}_{k+1}^{g(i_k^{z\star})} = \mathcal{S}_k^{g(i_k^{z\star})} \backslash \{i_k^{z\star}\}, \quad \forall \, z \in \mathcal{C}_k, \tag{8.5a}$$

$$\mathcal{S}_{k+1}^{\ell_k^{z\star}} = \mathcal{S}_k^{\ell_k^{z\star}} \cup \{i_k^{z\star}\}, \quad \forall \, z \in \mathcal{C}_k. \tag{8.5b}$$

Theorem 8.1 *If the initial partitioning $\mathcal{P}_0 = \{\mathcal{S}_0^1, \ldots, \mathcal{S}_0^m\}$ satisfies constraints (8.1b), and (8.1c), then these constraints are satisfied by $\mathcal{P}_k = \{\mathcal{S}_k^1, \ldots, \mathcal{S}_k^m\}$ for all $k \in \mathbb{Z}_{\geq 0}$ under the partitioning updating performed in (8.5).*

Proof It is assumed that $\mathcal{P}_0$ is a partition set such that $\bigcap_{\ell \in \mathcal{K}} \mathcal{S}_0^\ell = \emptyset$, and $\bigcup_{\ell \in \mathcal{K}} \mathcal{S}_0^\ell = \mathcal{S}$. Therefore, $\mathcal{S}_0^{g(i_0^{z\star})} \cap \mathcal{S}_0^{\ell_0^{z\star}} = \emptyset$ since $g(i_0^{z\star}) \neq \ell_0^{z\star}$ according to (8.4), and given that $g(i_0^{z\star}), \ell_0^{z\star} \in \mathcal{K}$, then

$$\left\{ \bigcup_{\ell \in \mathcal{K} \backslash \{g(i_0^{z\star}), \ell_0^{z\star}\}} \mathcal{S}_0^\ell \right\} \cup \left\{ \mathcal{S}_0^{g(i_0^{z\star})} \cup \mathcal{S}_0^{\ell_0^{z\star}} \right\} = \mathcal{S}. \text{ From (8.5), it is obtained that}$$

$$\mathcal{S}_{k+1}^{g(i_k^{z\star})} \cup \mathcal{S}_{k+1}^{\ell_k^{z\star}} = \left\{ \mathcal{S}_k^{g(i_k^{z\star})} \backslash \{i_k^{z\star}\} \right\} \cup \left\{ \{i_k^{z\star}\} \cup \mathcal{S}_k^{\ell_k^{z\star}} \right\},$$

$$= \mathcal{S}_k^{g(i_k^{z\star})} \cup \mathcal{S}_k^{\ell_k^{z\star}},$$

for all $z \in \mathcal{C}_k$. Finally, if $\mathcal{S}_k^{g(i_k^{z\star})} \cap \mathcal{S}_k^{\ell_k^{z\star}} = \emptyset$, for all $z \in \mathcal{C}_k$, then $\left\{ \mathcal{S}_k^{g(i_k^{z\star})} \backslash \mathcal{B} \right\} \cap \left\{ \mathcal{S}_k^{\ell_k^{z\star}} \cup \mathcal{B} \right\} = \emptyset$, for any set $\mathcal{B}$. Therefore, $\mathcal{S}_{k+1}^{g(i_k^{z\star})} \cap \mathcal{S}_{k+1}^{\ell_k^{z\star}} = \emptyset$. □

8.1.3 Partitioning in Flow-Based Distribution Systems

According to the algorithm for the m-partitioning problem, it is necessary to provide the number of partitions m, and the initial partition set $\mathcal{P}_0$, i.e., $\mathcal{S}_0^\ell$, for all $\ell \in \mathcal{K}$. This section is devoted to present the procedure to determine these two elements in the context of flow-based distribution systems. Many engineering systems may be modeled as a flow-based distribution system as presented in [12], e.g., water, energy or transportation systems. Section 7.1 has presented a simplified flow-based network since it only considers three elements. In general, flow-based distribution systems are composed of the following elements:

1. *Storage*: element that stores a resource and with both inflows and outflows. The set of storage elements is denoted by $\bar{\mathcal{S}}_{st}$. Notice that the notation corresponding to storage elements in the simplified flow-based network in Sect. 7.1 is different in order to distinguish the kind of network, i.e., simplified or general.
2. *Actuator*: element that manipulates the flow of the resource, having a unique inflow and outflow. The set of actuator elements is denoted by $\bar{\mathcal{S}}_{ac}$.
3. *Joint*: element without storage capabilities with an associated mass-balance constraint, and with both inflows and outflows. The set of joint elements is denoted by $\bar{\mathcal{S}}_{jo}$.
4. *Sink*: element that receives the resource from either a storage or joint element, e.g., demands in the system. The set of sink elements is denoted by $\bar{\mathcal{S}}_{si}$.
5. *Source*: element that provides the resource and with only outflows. The set of source elements is denoted by $\bar{\mathcal{S}}_{so}$.
6. *Flow*: directed link (i, j) allowing resource flow from an element $i \in \{\bar{\mathcal{S}}_{st} \cup \bar{\mathcal{S}}_{ac} \cup \bar{\mathcal{S}}_{so} \cup \bar{\mathcal{S}}_{jo}\}$ (storage, actuator, source, or joint elements) to an element $j \in \{\bar{\mathcal{S}}_{st} \cup \bar{\mathcal{S}}_{ac} \cup \bar{\mathcal{S}}_{si} \cup \bar{\mathcal{S}}_{jo}\}$ (storage, sink, or joint elements). The set of flow elements is denoted by $\bar{\mathcal{E}}$.

Let $\bar{\mathcal{G}} = (\bar{\mathcal{S}}, \bar{\mathcal{E}})$ be a directed graph representing a given flow-based distribution system describing the possible direction of the flows, where $\bar{\mathcal{S}} = \{\bar{\mathcal{S}}_{st} \cup \bar{\mathcal{S}}_{ac} \cup \bar{\mathcal{S}}_{si} \cup \bar{\mathcal{S}}_{so} \cup \bar{\mathcal{S}}_{jo}\}$ is the set of $r \in \mathbb{Z}_{>0}$ system elements, i.e., storage, actuator, sink, source, and joint elements. On the other hand, $\bar{\mathcal{E}} \subset \{(i, j) : i, j \in \bar{\mathcal{S}}\}$ is the set of flows from the element $i \in \bar{\mathcal{S}}$ to element $j \in \bar{\mathcal{S}}$.

Remark 8.3 Each physical element in the system has an equivalent node in the graph $\bar{\mathcal{G}}$. In addition, there is a direct relationship between the nodes $\bar{\mathcal{S}}$ in the flow graph $\bar{\mathcal{G}}$, and the nodes $\mathcal{S}$ in the information-sharing graph $\mathcal{G}$. ◇

The introduced elements for a flow-based distribution system, and the representation of the system by a directed graph, allow to identify some features of the system throughout indices and elements presented next.

Definition 8.1 (*Network resource-feeding index*) A non-source element $i \in \bar{\mathcal{S}} \backslash \bar{\mathcal{S}}_{so}$ in the flow-based distribution system has a resource-feeding index denoted by $\tau_i \leq |\bar{\mathcal{S}}_{so}|$, which gives information about the amount of source elements that may provide resource to the ith element, i.e., there are τ_i source elements that can feed the element $i \in \bar{\mathcal{S}} \backslash \bar{\mathcal{S}}_{so}$. ◇

Definition 8.2 (*Anchor elements*) A non-source element $i \in \bar{\mathcal{S}} \backslash \bar{\mathcal{S}}_{\text{so}}$ is anchor if there is only one source able to feed it, i.e., the resource-feeding index of an anchor element is unitary. Let $\tilde{\mathcal{A}} \subseteq \bar{\mathcal{S}}$ be the set of anchor elements within the flow-based network, i.e., $\tilde{\mathcal{A}} = \{i \in \bar{\mathcal{S}} \backslash \bar{\mathcal{S}}_{\text{so}} : \tau_i = 1\}$. ◇

Definition 8.3 (*Maximum resource-feeding index of the network*) The maximum resource-feeding index of the network is denoted by $\tau^\star$ and it corresponds to $\tau^\star = \max_{i \in \bar{\mathcal{S}}} \tau_i$. This index provides information about the non-source element in the system with more available source elements given by $i^\star_\tau \in \arg\max_{i \in \bar{\mathcal{S}}} \tau_i$. ◇

Definition 8.4 (*Maximum resource-feeding index per partition*) Given a partition of the flow-based distribution system into m sub-systems, the maximum resource-feeding index per partition is denoted by $\tau^\ell = \max_{i \in \bar{\mathcal{S}}^\ell} \tau_i$, for all $\ell \in \mathcal{K}$. Notice that source elements that do not belong to the current partition are also taken into account. Finally, the non-source element with more available source elements is given by $i^{\ell\star}_\tau \in \max_{i \in \bar{\mathcal{S}}^\ell} \tau_i$, for all $\ell \in \mathcal{K}$. ◇

Definition 8.5 (*Resource-feeding co-relation index*) The availability of resource at each partition $\ell \in \mathcal{K}$ is assessed with respect to the maximum resource index of the network given by the resource-feeding co-relation index denoted by $\varepsilon^\ell \in [0, 1]$. Formally, the resource-feeding co-relation index is $\varepsilon^\ell = \tau^\ell (\tau^\star)^{-1}$, for all $\ell \in \mathcal{K}$. Notice that if $\varepsilon^\ell = 1$ for a partition $\ell \in \mathcal{K}$, then the non-source element in the network with more available source elements belongs to the partition ℓ. ◇

Definition 8.6 (*Available source elements for non-source elements*) The available source elements that can provide resource to a non-source element is given by the set $\mathcal{R}_i$ and with $|\mathcal{R}_i| = \tau_i$ for all $i \in \bar{\mathcal{S}} \backslash \bar{\mathcal{S}}_{\text{so}}$. ◇

The number of partitions is determined by setting a desired minimum *resource-feeding co-relation index* (see Definition 8.5), i.e., partitions should satisfy that $\min_{\ell \in \mathcal{K}} \varepsilon^\ell \geq \varepsilon^\star$. In addition, further criteria to define the number of partitions may be included such that it is not desired that the elements $i^{\ell\star}_\tau$, for all $\ell \in \mathcal{K}$, were not neighbors. Once the aforementioned elements $i^{\ell\star}_\tau$, for all $\ell \in \mathcal{K}$ are identified in the graph $\bar{\mathcal{G}}$, then those can be associated to the corresponding variables in the information-sharing graph $\mathcal{G}$ and the initial partition $\mathcal{P}_0$ can be determined by adding each node from the set $\mathcal{S} \backslash \{i^{1\star}_\tau, \ldots, i^{m\star}_\tau\}$ to the partition associated to the nearest connected node from the set $\{i^{1\star}_\tau, \ldots, i^{m\star}_\tau\}$ (see Definition 8.3).

Figure 8.1 shows the summary of the partitioning procedure for a flow-based distribution system. First, a directed graph, denoted by $\bar{\mathcal{G}}$, is determined based on the flows throughout the system, which is used to determined the anchor elements, and the number of appropriate partitions that depend on the design parameter known as *resource-feeding co-relation index* (see Definition 8.5). On the other hand, an information-sharing graph, denoted by $\mathcal{G}$, describing how information dependence among the elements of the flow-based distribution system is determined according to a control strategy. An initial partition $\mathcal{P}_0$ is computed by using the number of partitions and the correspondence between elements in the directed graph and

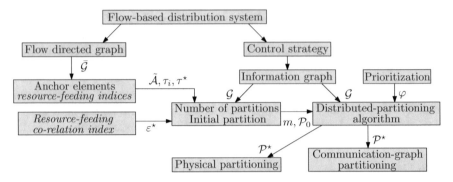

Fig. 8.1 Summary of the distributed partitioning procedure for flow-based distribution systems and for control purposes

the information-sharing graph, i.e., each variable considered in the controller corresponds to a physical element in the directed graph. Therefore, it follows to perform the distributed partitioning algorithm by setting a prioritization for the four objectives presented in Sect. 8.1.1. Finally, an optimal partitioning $\mathcal{P}^\star$ is obtained based on the information-sharing graph, and it can be interpreted into the physical system since there is a correspondence between each variable and an element in the flow-based distribution system.

8.2 Case Study: Partitioning of the BWSN

8.2.1 Management Criteria

This benchmark has been studied, for instance, in [13–17]. The same management criteria considered in Sect. 3.3.4 are taken into account in this chapter, i.e., the first objective is the minimization of economical costs, which are associated to the costs of the supplied water given by $\alpha_1^\top \mathbf{u}_k$, and the costs of the energy required to operate the system actuators (valves and pumps) given by $\alpha_{2,k}^\top \mathbf{u}_k$, where the vector $\alpha_2 \in \mathbb{R}^{n_u}$ varies along the time. The second objective is to operate the system smoothly to avoid possible damage, i.e., minimize $\|\Delta \mathbf{u}_k\|^2$, guaranteeing that the variations over the control inputs are smooth. Finally, it is desired that the volumes for all the tanks are higher than some predetermined constant volumes denoted by $\mathbf{x}_s \in \mathbb{R}^{n_x}$, i.e., constraint $\mathbf{x}_k \geq \mathbf{x}_s - \varsigma_k$ should be satisfied where $\varsigma_k \in \mathbb{R}^{n_x}_{\geq 0}$ is an auxiliary variable. The third objective is to minimize $\|\varsigma_{k+j}\|^2$. These three different objectives are prioritized with weights γ_1, γ_2, and γ_3 where $\gamma_1 > \gamma_3 > \gamma_2$.

8.2.2 Optimization Problem for the Predictive Controller

Once the system management criteria have been established with the three objectives, it can be set the optimization problem behind the MPC controller design as follows:

$$
\underset{\mathbf{u}_{k|k},\dots,\mathbf{u}_{k+H_p|k},\varsigma_{k|k},\dots,\varsigma_{k+H_p|k}}{\text{minimize}} \quad J(\mathbf{u},\varsigma) = \sum_{j=0}^{H_p-1} \left(\gamma_1 \left| \left(\alpha_1 + \alpha_{2,k+j} \right)^\top \mathbf{u}_{k+j} \right| + \gamma_2 \left\| \Delta \mathbf{u}_{k+j} \right\|^2 + \right.
$$

$$
\left. +\gamma_3 \left\| \varsigma_{k+j} \right\|^2 \right), \quad \text{(8.6a)}
$$

$$
\text{subject to} \quad \mathbf{x}_{k+i+1|k} = \mathbf{A}_d \mathbf{x}_{k+i|k} + \mathbf{B}_d \mathbf{u}_{k+i|k} + \mathbf{B}_l \mathbf{d}_{k+i|k}, \quad \text{(8.6b)}
$$

$$
0 = \mathbf{E}_u \mathbf{u}_{k+i|k} + \mathbf{E}_l \mathbf{d}_{k+i|k}, \quad \text{(8.6c)}
$$

$$
\mathbf{u}_{k+i|k} \in \mathcal{U}, \quad \text{(8.6d)}
$$

$$
\mathbf{x}_{k+i|k} \in \mathcal{X}, \quad \text{(8.6e)}
$$

$$
\mathbf{x}_{k+i|k} \geq \mathbf{x}_s - \varsigma_{k+i|k}, \quad \text{(8.6f)}
$$

$$
\varsigma_{k+i|k} \geq 0, \quad \text{(8.6g)}
$$

where (8.6b)–(8.6d) for all $i \in [0, H_p - 1] \cap \mathbb{Z}_{\geq 0}$, and (8.6e)–(8.6g) for all $i \in [0, H_p] \cap \mathbb{Z}_{\geq 0}$. Feasible sets $\mathcal{X}$, and $\mathcal{U}$ are defined as in Sect. 6.2, i.e., $\mathcal{X} \triangleq \left\{ \mathbf{x} \in \mathbb{R}^{n_x} : \underline{\mathbf{x}} \leq \mathbf{x} \leq \bar{\mathbf{x}} \right\}$, and $\mathcal{U} \triangleq \left\{ \mathbf{u} \in \mathbb{R}^{n_u} : \underline{\mathbf{u}} \leq \mathbf{u} \leq \bar{\mathbf{u}} \right\}$.

8.2.3 Computing the Information-Sharing Graph

The optimization problem in (8.6) may be written as a quadratic programing problem of the form [18]

$$
\underset{\xi_k}{\text{minimize}} \, \frac{1}{2} \xi_k^\top \mathbf{H} \xi_k + \mathbf{h}^\top \xi_k, \quad \text{(8.7a)}
$$

$$
\text{subject to} \, \mathbf{G} \xi_k = \mathbf{g}, \quad \text{(8.7b)}
$$

where $\xi_k = [\omega_k^\top \quad \dots \quad \omega_{k+H_p-1}^\top]^\top$, each element of ξ_k is $\omega_k = [\mathbf{u}_k^\top \quad \varsigma_k^\top \quad \mathbf{s}_k^\top]^\top$, and let $\mathbf{s}_k$ be the vector of slack variables for all the inequality constraints to make them equality constraints. Furthermore, the Lagrangian function corresponding to (8.7) is [19]

$$
L(\xi_k, \boldsymbol{\lambda}) = \frac{1}{2} \xi_k^\top \mathbf{H} \xi_k + \mathbf{h}^\top \xi_k + \boldsymbol{\lambda}^\top \left(\mathbf{G} \xi_k - \mathbf{g} \right).
$$

Therefore, the KKT conditions are obtained from $\nabla_{\xi_k} L(\xi_k^\star, \boldsymbol{\lambda}^\star) = \mathbf{0}$, and $\nabla_{\boldsymbol{\lambda}} L(\xi_k^\star, \boldsymbol{\lambda}^\star) = \mathbf{0}$ giving by the following equality:

$$\underbrace{\begin{bmatrix} \mathbf{H} & \mathbf{G}^\top \\ \mathbf{G} & \mathbf{0} \end{bmatrix}}_{\mathbf{\Psi}} \begin{bmatrix} \xi_k^\star \\ \lambda^\star \end{bmatrix} = \begin{bmatrix} -\mathbf{h} \\ \mathbf{g} \end{bmatrix}. \tag{8.8}$$

Notice that the matrix $\mathbf{\Psi}$ represents the information dependence among variables ξ_k and λ in order to solve in (8.8) and thus solve the optimization problem (8.6). In this regard, if there is a node per each variable in (8.8), then the adjacency matrix $\mathbf{A}$ of the graph $\mathcal{G}$ is given by $a_{ij} = 1$ if $\psi_{ij} \neq 0$, and $a_{ij} = 0$, otherwise, defining the topology of the information-sharing graph.

Remark 8.4 The information dependence among variables, which generates the infor-mation-sharing graph $\mathcal{G}$, can be computed considering any control strategy and for any system. In this regard, the proposed partitioning method is general.   ◇

8.2.4 Anchor Elements

In order to find the anchor elements in the BWSN, the directed graph $\bar{\mathcal{G}} = (\bar{\mathcal{S}}, \bar{\mathcal{E}})$ is obtained by replacing each element of the network (i.e., sources, actuators, tanks, sinks, and mass-balance joints) for nodes of the graph denoted by $\bar{\mathcal{S}}$ and replacing each flow for a graph edge denoted by $\bar{\mathcal{E}}$. The *network resource-feeding indices* are shown in Table 8.1 (see Definition 8.1). Therefore the anchor-storage elements from the $i \in \tilde{\mathcal{A}}$ are x_1, x_2, x_5, and x_8 (see Definition 8.2). Moreover, these storage elements and their respective source elements must belong to the same partition, e.g., the tank x_1, and the source s_2 should belong to the same partition. Finally, the *maximum resource-feeding index of the network* is given by $\tau^\star = 7$.

The appropriate number of partitions is determined based on the *resource-feeding co-relation index*, and further conditions may be added to this criterion. As an example, suppose that it is desired to have a *resource-feeding co-relation index* given by $\varepsilon^\star = 0.25$ (i.e., it is desired that the partition with less-available source elements has at least 25% of the maximum resource elements that can provide resource in the system $\tau^\star$). Additionally, it is desired that the storage elements corresponding to the

Table 8.1 Resource-feeding network indices for the storage elements in the BDWN

Storage element $i \in \bar{\mathcal{S}}_{\text{st}}$	Resource-feeding indices τ_i	Source element $\mathcal{R}_i$
x_1	1	s_2
x_2	1	s_1
x_5, x_8	1	s_6
x_6	2	s_1, s_6
$x_{14}, \ldots, x_{16}$	2	s_7, s_8
$x_7, x_9, \ldots, x_{11}, x_{13}$	5	$s_1, s_3, s_6, \ldots, s_8$
x_3, x_4, x_{12}, x_{17}	7	$s_1, s_3, s_4, \ldots, s_8$

maximum resource-feeding index per partition $i_\tau^{\ell\star}$, for all $\ell \in \mathcal{K}$, do not have the same source elements $\mathcal{R}_i$. Therefore, x_3, x_4, x_{12}, and x_{17} should belong to the same partition since they have a common set of available sources (first partition with *resource-feeding co-relation index* $\varepsilon^1 = 1$). The same situation happens with x_7, x_9, x_{10}, x_{11}, and x_{13} (second partition with *relation resource-feeding index* $\varepsilon^2 = 0.7142$); x_{14}, x_{15}, and x_{16} (third partition with *resource-feeding co-relation index* $\varepsilon^3 = 0.2857$); and x_6 (fourth partition with *resource-feeding co-relation index* $\varepsilon^4 = 0.2857$).

8.3 DMPC Applying Large-Scale System Partitioning

Once the appropriate number of partitions is determined by using a desired *resource-feeding co-relation index* $\varepsilon^\star$, then the partitioning algorithm is run. In this chapter, it is proposed to use the information-sharing graph $\mathcal{G}$ computed in Sect. 8.2.3, which is shown in Fig. 8.2. The initial partition $\mathcal{P}_0$ considered within the algorithm is determined by using the nodes associated to the elements corresponding to the *maximum partition resource-feeding indices* $i_\tau^{\ell\star}$, for all $\ell \in \mathcal{K}$. Therefore, the rest of nodes are incorporated to the closest partition with which there is connection. The partitioning algorithm is performed with weights $\varphi = [0.5 \quad 0.2 \quad 0.2 \quad 1]^\top$, and with the

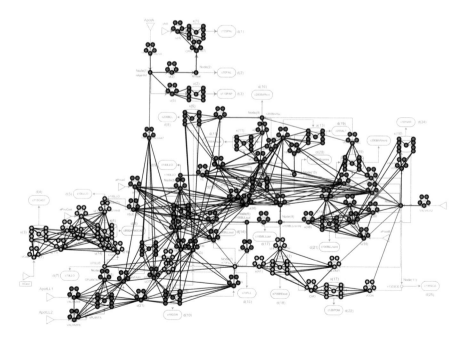

Fig. 8.2 Information-sharing graph of the BWSN. The optimal partitioning is also presented (partitions 1–4 with colors green, blue, magenta, and red, respectively). Links within the same partition with black color and links connecting different partitions with red color

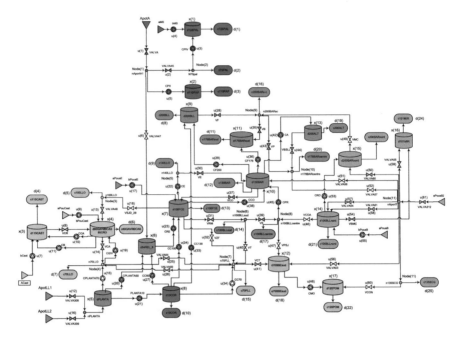

Fig. 8.3 Aggregate model of the BSWN. The optimal physical partitioning is also presented (sub-systems 1–4 with colors green, blue, magenta, and red, respectively)

parameter $\varkappa = 0$ for the algorithm ending-up condition. The optimal partitioning $\mathcal{P}^\star$ is the one presented in Fig. 8.2. It is important to highlight that the total number of communication links in order to compute the optimal control input according to problem (8.8) is $(\mathbb{1}^\top A \mathbb{1})/2 = 361$. Furthermore, the optimal partition $\mathcal{P}^\star$ has 13 links among partitions that is the 3.6% of the total number of communication links, representing reduced communication dependence among different partitions, which is desired for the design of non-centralized controllers.

With the optimal partition $\mathcal{P}^\star$ presented in Fig. 8.2, the information-sharing graph is interpreted/translated into the physical system, obtaining the physical partitioning into m sub-systems presented in Fig. 8.3 (the indices of the m sub-systems are given by the set $\mathcal{K}$). With the m-partitioning, an LMPC controller is designed for each sub-system, identifying the information dependence among them as in [2]. This procedure results in the sub-system dependence presented in Fig. 8.4 (partitions from 1 to m correspond to colors green, blue, magenta, and red, respectively), where the terms $\mu_{r\ell}$ are the information provided by sub-system $r \in \mathcal{K}$ to sub-system $\ell \in \mathcal{K}$. It can be seen that the optimization problem behind the LMPC controller associated to the Sub-system 1 may be solved since it does not require information from other sub-systems. After that, Sub-system 1 provides information μ_{12}, and μ_{13} to Sub-system 2 and 3, respectively. Therefore, the optimization problem behind the LMPC associated to the Sub-system 2 is solved, providing information μ_{23}, and μ_{24} to

Fig. 8.4 Hierarchical and information dependence among the m sub-systems

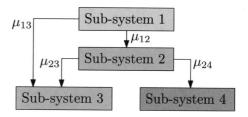

Sub-systems 3 and 4, respectively. The mentioned information shared among sub-systems is structured as follows:

$$\mu_{12} = [\tilde{\mathbf{u}}_{15,k}^{\star\top} \quad \tilde{\mathbf{u}}_{18,k}^{\star\top} \quad \tilde{\mathbf{u}}_{32,k}^{\star\top} \quad \tilde{\mathbf{u}}_{34,k}^{\star\top} \quad \tilde{\mathbf{u}}_{40,k}^{\star\top} \quad \tilde{\mathbf{u}}_{47,k}^{\star\top}]^{\top},$$

$$\mu_{13} = [\tilde{\mathbf{u}}_{56,k}^{\star\top} \quad \tilde{\mathbf{u}}_{60,k}^{\star\top}]^{\top},$$

$$\mu_{23} = [\tilde{\mathbf{u}}_{46,k}^{\star\top} \quad \tilde{\mathbf{u}}_{49,k}^{\star\top}]^{\top},$$

$$\mu_{24} = \tilde{\mathbf{u}}_{6,k}^{\star},$$

where $\tilde{\mathbf{u}}_{i,k}^{\star} = [u_{i,k}^{\star} \quad \cdots \quad u_{i,k+H_p-1}^{\star}]^{\top} \in \mathbb{R}^{H_p}$. Different from the work presented in [2], there are not cycles in the hierarchical structure describing information dependence among sub-systems, therefore, it is not required to solve a CSP.

In order to evaluate the performance of the DMPC controller, a comparison with the performance of a CMPC is made. Both approaches are designed with the same prioritization weights, i.e., $\gamma_1 = 1$, $\gamma_2 = 0.001$, and $\gamma_3 = 0.1$, and simulations are made for four days. Table 8.2 shows the costs for each day denoted by y. The operational costs of energy

$$\text{KPI}_E(\text{day}) = \sum_{k=24\text{day}-24}^{24\text{day}} \alpha_1^{\top} \mathbf{u}_k,$$

the operational costs of water

Table 8.2 Costs comparison between the CMPC and DMPC controllers

Day	CMPC			DMPC		
	Energy $\text{KPI}_E(\text{day})$	Water $\text{KPI}_W(\text{day})$	Slew rate $\text{KPI}_{\Delta u}(\text{day})$	Energy $\text{KPI}_E(\text{day})$	Water $\text{KPI}_W(\text{day})$	Slew rate $\text{KPI}_{\Delta u}(\text{day})$
1	9.8133	5.8964	0.23766	10.499	2.6238	0.35532
2	8.5959	5.8829	0.018947	10.266	2.4903	0.010012
3	8.5959	5.8829	0.018947	10.267	2.489	0.010074
4	8.5959	5.8829	0.018947	10.267	2.4886	0.010066
Total	52.793	35.311	0.33239	61.833	15.069	0.4056

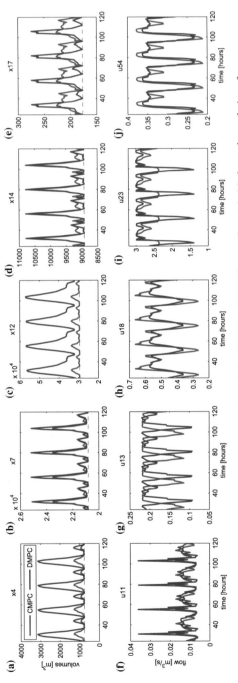

Fig. 8.5 Evolution of five system states, and five control inputs for both the CMPC and DMPC controllers. (**a**)–(**e**) shows the evolution of x_4, x_7, x_{12}, x_{14}, and x_{17}, respectively. (**f**)–(**j**) shows the evolution of u_{11}, u_{13}, u_{18}, u_{23}, and u_{54}, respectively

$$\text{KPI}_W(\text{day}) \;=\; \sum_{k=24\text{day}-24}^{24\text{day}} \alpha_{2,k}^\top \mathbf{u}_k,$$

and costs associated to the smooth operation (slew rate)

$$\text{KPI}_{\Delta u}(\text{day}) \;=\; \sum_{k=24\text{day}-24}^{24\text{day}} \|\Delta \mathbf{u}_k\|^2.$$

It can be seen that, the system operation with the distributed control approach implies higher energy costs and also more variations over the control signals in comparison to the centralized-control scheme. In contrast, the costs associated to the water are lower with the distributed approach.

Figure 8.5 shows the evolution of five system states and five control inputs for both the CMPC, and DMPC controllers. It can be seen that the periodicity of the signals for both cases is the same. In general, the performance corresponding to the DMPC controller exhibits higher variations over the evolution of the system states. In particular, Fig. 8.5a, c show higher amplitudes in the oscillation of the system states x_4, and x_{12}, respectively. Moreover, the same occurs with the control inputs, e.g., Fig. 8.5i shows a higher amplitude for the control input u_{23}. However, the computational tasks of the CMPC controller are divided and assigned to four different LMPC controllers, and the distributed controller continues achieving the control objectives and satisfying all the constraints.

8.4 Summary

A multi-objective partitioning procedure considering several aspects such as amount of links connecting different partitions, size of partitions, distance among elements, and importance of links has been presented in order to determine the appropriate partitioning in a large-scale flow-based system. As one of the most relevant features, the proposed partitioning is a general methodology since it can be implemented for any dynamical system, including flow-based network systems, and any control strategy since an information sharing graph is considered.

Motivated by the fact that some nominal conditions might vary along the time for a dynamical system, e.g., disturbances affecting the system under control, next chapter in this thesis extends the partitioning procedure to the dynamical case. Thus, the large-scale system can be divided conveniently and permanently in order to enhance the performance of closed-loop systems operated by DMPC controllers.

References

1. Barreiro-Gomez J, Ocampo-Martinez C, Quijano N, Maestre JM (2017a) Non-centralized control for flow-based distribution networks: a game-theoretical insight. J Frankl Inst 354(14):5771–5796
2. Ocampo-Martinez C, Bovo S, Puig V (2011) Partitioning approach oriented to the decentralised predictive control of large-scale systems. J Process Control 21(2011):775–786
3. Sezer ME, Šiljak DD (1986) Nested $\varepsilon-$decompositions and clustering of complex systems. Automatica 22(3):321–331
4. Chandan V, Alleyne A (2013) Optimal partitioning for the decentralized thermal control of buildings. IEEE Trans Control Syst Technol 21(5):1756–1770
5. Kleinberg MR, Miu K, Segal N, Lehmann H, Figura TR (2014) A partitioning method for distributed capacitor control of electric power distribution systems. IEEE Trans Power Syst 29(2):637–644
6. Nayeripour M, Fallahzadeh-Abarghouei H, Waffenschmidt E, Hasanvand S (2016) Coordinated online voltage management of distributed generationusing network partitioning. Electr Power Syst Res 141(2016):202–209
7. Xie L, Cai X, Chen J, Su H (2016) GA based decomposition of large scale distributed model predictive control systems. Control Eng Pract 57(2016):111–125
8. Kamelian S, Salahshoor K (2015) A novel graph-based partitioning algorithm for large-scale dynamical systems. Int J Syst Sci 46(2):227–245
9. Núñez A, Ocampo-Martinez C, Maestre JM, De Schutter B (2015) Time-varying scheme for noncentralized model predictive control of large-scale systems. Math Probl Eng 2015(560702):1–17
10. Gupta A (1997) Fast and effective algorithms for graph partitioning and sparse-matrix ordering. IBM J Res Dev 41(1):171–183
11. Olfati-Saber R, Fax JA, Murray RM (2007) Consensus and cooperation in networked multi-agent systems. Proc IEEE 95(1):215–233
12. Grosso J (2015) Economic and robust operation of generalised flow-based networks. Doctoral dissertation. Automatic Control Department. Universidad Politècnica de Catalunya,
13. Barreiro-Gomez J, Ocampo-Martinez C, Quijano N (2017) Dynamical tuning for mpc using population games: a water supply network application. ISA Trans 69(2017):175–186
14. Barreiro-Gomez J, Ocampo-Martinez C, Quijano N (2017c) Partitioning for large-scale systems: a sequential dmpc design. In: Proceedings of the 20th IFAC world congress. Toulouse, France, pp 8838–8843
15. Barreiro-Gomez J, Quijano N, Ocampo-Martinez C (2016a) Distributed MPC with time-varying communication network: a density-dependent population games approach. In: Proceedings of the 55th IEEE Conference on decision and control (CDC). Las Vegas, USA, pp 6068–6073
16. Barreiro-Gomez J, Ocampo-Martinez C, Quijano N (2015) Evolutionary-game-based dynamical tuning for multi-objective model predictive control. In: Olaru S, Grancharova A, Lobo Pereira F (eds) Developments in model-based optimization and control. Springer, Berlin, pp 115–138
17. Barreiro-Gomez J, Quijano N, Ocampo-Martinez C (2016b) Constrained distributed optimization: a population dynamics approach. Automatica 69:101–116
18. Maciejowski J (2002) Predictive control: with constraints. Pearson education
19. Chong EKP, Zak SH (2013) An introduction to optimization. Wiley series in discrete mathematics and optimization. Wiley, New York

Chapter 9
Distributed System Partitioning and DMPC

This chapter addresses the design of a DMPC based on both DDPG [1] and a distributed dynamical system partitioning [2–4]. To this end, the contributions presented in Chaps. 6 and 8 are combined to design a distributed optimization-based controller also considering a dynamical system partitioned. Depending on the current system states, some constraints are neglected in order to reduce the number of decision variables of the optimization problem behind the MPC controller design. Thus, the size of the information-sharing network is also reduced. The partitioning algorithm is performed to determine the appropriate set of sub-systems in function of the information-sharing network [2]. Finally, the DDPG approach computes all the optimal control inputs at each time instant [1], taking advantage of the population dynamics characteristics as studied in [5–7]. Notice that, due to the fact that the information-sharing network varies along the time, then the obtained optimal system partitioning is also different.

First, the general QP problem presented in Chap. 6 is recalled, where a DMPC without system partitioning is presented. Then, the partitioning presented in Chap. 8, where a sequential DMPC controller has been designed, is used in this chapter as an off-line system partitioning for the design of a parallel DMPC design based on population games. The consideration that the links connecting different partitions are not available for all the time is taken into account. Once this static system partitioning is presented for the control design, then the partitioning is applied in a dynamical manner preserving the same population-games strategy.

9.1 Control Problem Statement

Consider the same MPC controller design presented in Chap. 6, and with an optimization problem as in (6.17), i.e.,

$$\underset{\mathbf{U}_k}{\text{minimize}} \quad \mathbf{U}_k^\top \boldsymbol{\Phi} \, \mathbf{U}_k + \boldsymbol{\phi}_k^\top \mathbf{U}_k, \tag{9.1}$$

© Springer International Publishing AG, part of Springer Nature 2019
J. Barreiro-Gomez, *The Role of Population Games in the Design of Optimization-Based Controllers*, Springer Theses,
https://doi.org/10.1007/978-3-319-92204-1_9

subject to

$$\mathbf{E}\,\mathbf{U}_k \leq \mathbf{e}_k,$$ (9.2)

$$\mathbf{G}\,\mathbf{U}_k = \mathbf{g}_k.$$ (9.3)

The cost function is re-written, and the constraints can be compacted by adding slack variables, i.e.,

$$\underset{\xi}{\text{minimize}} \quad \underbrace{\begin{bmatrix} \mathbf{U}_k^\top & \mathbf{s}^\top \end{bmatrix}}_{\xi_k^\top} \underbrace{\begin{bmatrix} \boldsymbol{\Phi} & \mathbf{0}_{n_u \times q} \\ \mathbf{0}_{q \times n_u} & \mathbf{0}_{q \times q} \end{bmatrix}}_{\boldsymbol{\Psi}} \underbrace{\begin{bmatrix} \mathbf{U}_k \\ \mathbf{s} \end{bmatrix}}_{\xi_k} + \underbrace{\begin{bmatrix} \boldsymbol{\phi}_k^\top & \mathbf{0}_q^\top \end{bmatrix}}_{\psi_k^\top} \underbrace{\begin{bmatrix} \mathbf{U}_k \\ \mathbf{s} \end{bmatrix}}_{\xi_k},$$ (9.4)

subject to

$$\underbrace{\begin{bmatrix} \mathbf{E} & \mathbb{I}_q \\ \mathbf{G} & \mathbf{0}_{r \times q} \end{bmatrix}}_{\mathbf{H}} \underbrace{\begin{bmatrix} \mathbf{U}_k \\ \mathbf{s} \end{bmatrix}}_{\xi_k} = \underbrace{\begin{bmatrix} \mathbf{e}_k \\ \mathbf{g}_k \end{bmatrix}}_{\mathbf{h}_k}.$$ (9.5)

Therefore, the optimization problem behind the MPC controller is formulated in the form (6.11), and it can be solved in a distributed manner by using the D3RD in (6.2), the D3SD in (6.4), or the D3PD in (6.6), as explained in Sect. 6.1.

9.2 System Partitioning for DDPG-Based DMPC Control Design

This section presents how to apply a static off-line system partitioning for the design of a DMPC controller based on DDPG. The first step in order to apply the proposed distributed system partitioning algorithm is to determine the information-sharing graph. The adjacency matrix that determines the graph $\mathcal{G}$ is given by the required information-sharing matrix, i.e., $\mathbf{A} = \tilde{\boldsymbol{\Theta}}$. Recalling the optimization problem behind the MPC controller in (9.4), the corresponding Lagrangian function [8] is as follows:

$$L(\xi_k, \lambda_k) = \xi_k^\top \boldsymbol{\Psi} \xi_k + \psi_k \xi_k + \boldsymbol{\lambda}^\top \left(\mathbf{H} \xi_k - \mathbf{h}_k \right).$$ (9.6)

Therefore, the Karush-Kuhn-Tucker conditions are obtained from $\nabla_{\xi_k} L(\xi_k^\star, \lambda_k^\star) = \mathbf{0}$, and $\nabla_{\lambda_k} L(\xi_k^\star, \lambda_k^\star) = \mathbf{0}$. When the information-sharing graph $\mathcal{G}$ with adjacency matrix $\mathbf{A}$ is available, i.e., when there is not partitioning of the system, then the fitness functions are given by the following expression corresponding to (6.14):

$$\mathbf{f}(\mathbf{p}) = \underbrace{\begin{bmatrix} 2\boldsymbol{\Psi} & \mathbf{H}^{\top} \\ -\mathbf{H} & \mathbf{0}_{(q+r)\times(q+r)} \end{bmatrix}}_{\boldsymbol{\Omega}} \underbrace{\begin{bmatrix} \boldsymbol{\xi}_{k}^{\star} \\ \boldsymbol{\lambda}_{k}^{\star} \end{bmatrix}}_{\mathbf{p}^{\star}} + \begin{bmatrix} \boldsymbol{\psi}_{k} \\ \mathbf{h}_{k} \end{bmatrix}.$$

Moreover, when the system is partitioned, there are two classifications for the informa-tion-sharing links. Notice that when the partitioning is performed, then there are information-sharing links within the same partition and links connecting different partitions. According to the partitioning approach presented in Sect. 8.1.2, an optimal partition $\mathcal{P}^{\star}$ has an associated graph (not necessarily connected) denoted by $\tilde{\mathcal{G}}$. The graph $\tilde{\mathcal{G}}$ represents the graph whose links $\tilde{\mathcal{E}}$ correspond to the links connecting different partitions and with adjacency matrix $\tilde{\mathbf{A}}$.

9.2.1 DMPC Controller with DDPG and Static System Partitioning

The optimization problem behind the DMPC controller is solved by capturing information from other partitions throughout the information-sharing links $\tilde{\mathcal{E}}$. Afterwards, these links $\tilde{\mathcal{E}}$ are disconnected, and the DDPG evolve independently at each partition satisfying the information-sharing graph now imposed with an adjacency matrix $\hat{\mathbf{A}}$ as presented in (9.7). Therefore, the amount of required information-sharing links along the time is reduced since the links $\tilde{\mathcal{E}}$ are not needed for all the time. Hence, the fitness functions are computed as follows:

$$\mathbf{f}(\mathbf{p}) = \Big[\underbrace{\big(\mathbf{A} - \tilde{\mathbf{A}}\big)}_{\hat{\mathbf{A}}} \circ \boldsymbol{\Omega} \Big] \underbrace{\begin{bmatrix} \boldsymbol{\xi}_{k} \\ \boldsymbol{\lambda}_{k} \end{bmatrix}}_{\mathbf{p}} + \Big[\tilde{\mathbf{A}} \circ \boldsymbol{\Omega} \Big] \begin{bmatrix} \boldsymbol{\xi}_{k-1}^{\star} \\ \boldsymbol{\lambda}_{k-1}^{\star} \end{bmatrix} + \begin{bmatrix} \boldsymbol{\psi}_{k} \\ \mathbf{h}_{k} \end{bmatrix}. \tag{9.7}$$

Remark 9.1 Notice that the fitness functions in (9.7) do not share information among different partitions since the DDPG evolve satisfying an information-sharing graph whose adjacency matrix is $\hat{\mathbf{A}} = \mathbf{A} - \tilde{\mathbf{A}}$. ◇

Once the equilibrium point $\mathbf{p}^{\star}$ for the DDPG is obtained, the procedure can be repeated in order to find a better solution for the optimization problem, i.e., the values corresponding to the information provided throughout the links $\tilde{\mathcal{E}}$ are updated. In this regard, information from other partitions can be updated to improve the solution, and in fact, the same solution that is gotten without performing any partitioning can be obtained. To this end, the update of information from other partitions can be repeated until $[\boldsymbol{\xi}_{k}^{\star\top} \ \boldsymbol{\lambda}_{k}^{\star\top}]^{\top} = [\boldsymbol{\xi}_{k-1}^{\star\top} \ \boldsymbol{\lambda}_{k-1}^{\star\top}]^{\top}$. Moreover, in order to avoid infeasibility in the DMPC controller based on DDPG with a system partitioning, more importance is assigned to those links associated to the equality constraints, throughout the relevance matrix $\mathbf{C}$ as presented in the partitioning algorithm (see Sect. 8.1.2). Thus, the optimal partitioning $\mathcal{P}^{\star}$ integrates those links (critical for feasibility) within partitions avoiding that they belong to $\tilde{\mathcal{E}}$.

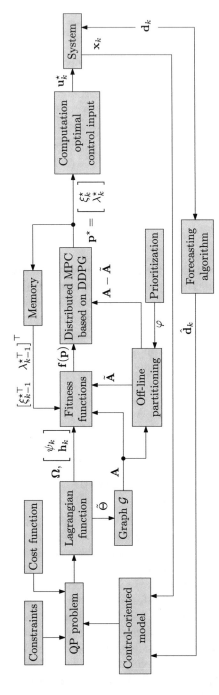

Fig. 9.1 Summary of the DMPC controller with distributed DDPG and static system partitioning

Figure 9.1 presents the summary scheme of the DMPC controller with DDPG and with static system partitioning. The Lagrangian function allows to compute the informa-tion-sharing graph $\mathcal{G}$, and provides information about $\boldsymbol{\Omega}$, and $[\boldsymbol{\psi}_k^\top \ \mathbf{h}_k^\top]^\top$. On the other hand, an off-line partitioning procedure is performed, providing the constant adjacency matrix of the graph $\tilde{\mathcal{G}}$, i.e., $\tilde{\mathbf{A}}$. With data from both the Lagrangian function and the off-line partitioning, the fitness functions are computed as in (9.7). Thus, the DMPC controller based on DDPG computes the optimal control input that is applied to the system.

9.2.2 DMPC Controller with DDPG and Dynamical System Partitioning

The system partitioning approach presented in Sect. 8.1 considers several aspects. One of the objectives dictates that all the partitions should have similar size, which results in having DMPC controllers with similar number of decision variables and constraints. Therefore, this aspect implies to have similar computational burden for all the distributed controllers.

Moreover, as discussed in Sect. 9.2.1, the relevance matrix $\mathbf{C}$ can be selected appropriately by assigning higher weights to those links involved in equality constraints in order to avoid infeasibility for the optimization problem. On the other hand, Sect. 6.2 addresses the case in which the set of considered constraints varies depending on the current state of the system, reducing conveniently the number of required information-sharing links in order to compute the optimal control input at each time instant.

Consider that the information-sharing graph $\mathcal{G}$ varies over time as it has been presented in Sect. 6.2, i.e., that the adjacency matrix $\mathbf{A}$ varies along the time. Then, due to the fact that the information-sharing graph varies, it is necessary to determine the appropriate partitioning of the system in a dynamical manner such that the partitioning criteria hold. Besides, the set of links $\tilde{\mathcal{E}}$ also varies along the time, which are links that are not required for all the time but only to update the fitness functions, i.e.,

$$\mathbf{f}(\mathbf{p}) = \left[\underbrace{\left(\mathbf{A}_k - \tilde{\mathbf{A}}_k \right) \circ \boldsymbol{\Omega}}_{\hat{\mathbf{A}}_k} \right] \underbrace{\begin{bmatrix} \boldsymbol{\xi}_k \\ \boldsymbol{\lambda}_k \end{bmatrix}}_{\mathbf{p}} + \left[\tilde{\mathbf{A}}_k \circ \boldsymbol{\Omega} \right] \begin{bmatrix} \boldsymbol{\xi}_{k-1}^\star \\ \boldsymbol{\lambda}_{k-1}^\star \end{bmatrix} + \begin{bmatrix} \boldsymbol{\psi}_k \\ \mathbf{h}_k \end{bmatrix}. \tag{9.8}$$

Remark 9.2 Notice that the fitness functions in (9.8) do not share information among different partitions since the DDPG evolve satisfying an information-sharing graph whose adjacency matrix is $\hat{\mathbf{A}}_k = \mathbf{A}_k - \tilde{\mathbf{A}}_k$. ◇

The application of a dynamical system partitioning for the design of distributed controllers has several advantages, e.g., the computational burden is distributed for all the non-centralized controllers for all the sub-systems. When considering that the information-sharing graph varies along the time getting rid of non-active constraints,

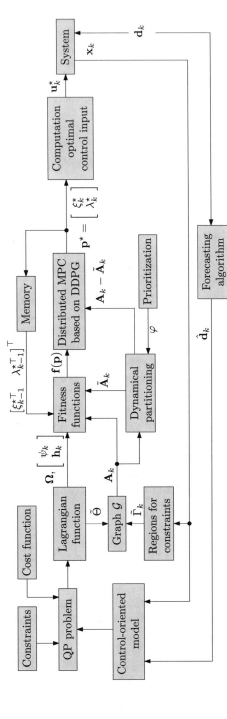

Fig. 9.2 Summary of the DMPC controller with distributed DDPG and dynamical system partitioning

then the computational burden associated to the optimization problem is reduced. The same partitioning objectives are still considered although the information-sharing graph varies along the time, and the appropriate set of not required links $\tilde{\mathcal{E}}$ is found after performing the system partitioning. Finally, suppose that there is any inconvenient or fault at any partition. For the conventional CMPC approach, the whole system is affected. In contrast, when adopting a non-centralized MPC controller approach using applying a system partitioning, any inconvenient at a partition is decoupled from other.

Figure 9.2 shows the summary of the DMPC with DDPG and with dynamical system partitioning. The current state of the system determines the *non-safe* regions for the constraints as in Sect. 6.2, defining the adjacency matrix $\mathbf{A}_k$ associated to the information-sharing graph. Therefore, the system partitioning is performed with the adjacency matrix $\mathbf{A}$, finding the appropriate matrix $\tilde{\mathbf{A}}$ describing all the links connecting different partitions. It follows that the fitness functions can be defined by using the Lagrangian function and matrices $\mathbf{A}$, and $\tilde{\mathbf{A}}$, and the DMPC controller with DDPG computes the optimal control input $\mathbf{u}_k^\star$. Afterwards, the new system state is measured and the aforementioned procedure is repeated, i.e., it is found the new information-sharing graph with matrix $\mathbf{A}$, the new partitioning with associated matrix $\tilde{\mathbf{A}}$ is computed, new fitness functions $\mathbf{f}(\mathbf{p})$ are established, and a new optimal control input is computed at each time instant in a distributed manner.

9.3 Case Study: Barcelona Water Supply Network

Consider the same case study presented in Sect. 3.3.3, i.e., the BWSN presented in Fig. 3.4. This benchmark has been studied, for instance, in [1, 2, 7, 9, 10]. It is proposed the design of three DMPC controllers based on DDPG for three different scenarios. Then, the DMPC is compared with a CMPC that uses a constant information-sharing graph without partitioning. The different scenarios are described as follows:

- *Scenario 1*: DMPC controller based on DDPG with constant information-sharing network and static system partitioning for the BWSN.
- *Scenario 2*: DMPC controller based on DDPG with time-varying information-sharing network and dynamical system partitioning for the BWSN.
- *Scenario 3*: CMPC controller based on DDPG with constant information-sharing network for the BWSN.

In order to evaluate the performance of the three scenarios, two different KPIs are proposed as in Chap. 6, i.e., the same economical costs $\text{KPI}_{\text{costs}}(\text{day})$ presented in (6.22) and communication costs $\text{KPI}_{\text{costs}}(\text{day})$ presented in (6.23), where $\mathbf{M}_k = \mathbf{A} - \tilde{\mathbf{A}}$ for Scenario 1 (constant), and $\mathbf{M}_k = \mathbf{A}_k - \tilde{\mathbf{A}}_k$ for Scenario 2, and $\mathbf{M}_k = \mathbf{A}$ for Scenario 3 (constant).

Regarding the Scenario 1, the same parameters for the partitioning algorithm of the Chap. 8 are used for this scenario, i.e., the information-sharing network and the

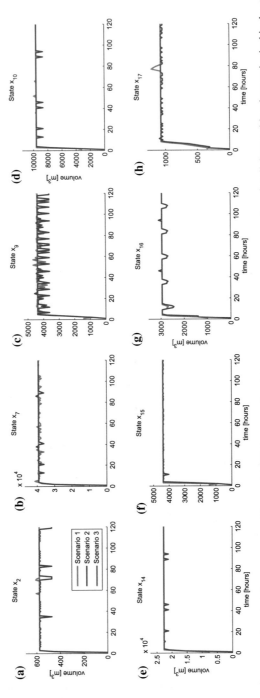

Fig. 9.3 Evolution of eight system states. Figures (a)–(h) correspond to states x_2, x_7, x_9, x_{10}, x_{14}, x_{15}, x_{16}, and x_{17} for all the considered scenarios in this chapter

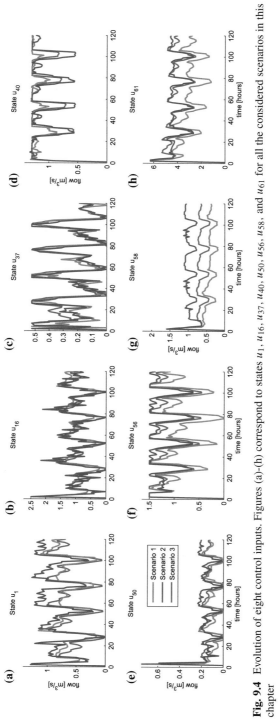

Fig. 9.4 Evolution of eight control inputs. Figures (a)–(h) correspond to states u_1, u_{16}, u_{37}, u_{40}, u_{50}, u_{56}, u_{58}, and u_{61} for all the considered scenarios in this chapter

Fig. 9.5 Evolution of the connected links for the three scenarios

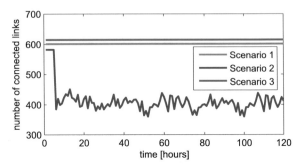

system partitioning presented in Fig. 8.2 are used for the DMPC controller based in DDPG. Figure 9.3 shows the evolution of some states achieving the imposed reference and control inputs for the DMPC controller with DDPG and the static system partitioning presented in Fig. 8.2a. Moreover, Fig. 9.4 presents the evolution of the corresponding variables with the CMPC controller without system partitioning. It can be seen that the behavior is the same for the evolution of states since the differences between the control inputs is quite subtle. Finally, Fig. 9.5 shows the constant number of connected links in the information-sharing network along the time.

On the other hand, the simulation parameters are selected as in Scenario 1, i.e., the reference has been selected to be $x_r = 0.6\bar{x}$, and the weights in the cost function in (8.6a) are selected to be $\tilde{Q} = \mathbb{I}_{n_x}$, $\tilde{R} = 1000\mathbb{I}_{n_u}$, and $\gamma = 1$. The partitioning algorithm is performed with weights $\varphi = [1 \quad 0.26 \quad 0.1 \quad 0.5]^\top$, and with the parameter $\varkappa = 0$ for the algorithm end-up condition.

Figures 9.6, and 9.7 present the time-varying information-sharing graphs, the corresponding optimal system partitioning highlighting the links connecting different partitions, and the corresponding physical partitioning for the BWSN.[1] Figure 9.6a corresponds to a unique time instant, i.e., $k = 15$, whereas Fig. 9.6b corresponds to 28 time instants, being the most frequent partitioning. Moreover, Fig. 9.7a corresponds to four time instants, i.e., $k \in \{43, 67, 91, 115\}$, and Fig. 9.7b corresponds to nine time instants, i.e., $k \in \{55, \ldots, 47, 79, \ldots, 81, 103, \ldots, 105\}$.

Looking at Figs. 9.3 and 9.4 again, Fig. 9.3 shows the evolution of some states and Fig. 9.4 shows the evolution of some control inputs for the DMPC controller with DDPG and the dynamical system partitioning in comparison to the evolution of the respective variables for the same controller without system partitioning. It can be seen that the system states achieve the reference. Making a comparison with respect to the performance observed for the case with static system partitioning (see Scenario 2), the dynamical system partitioning produces more variations around the reference (even more than the ones generated by the permanent disturbances in the system). This behavior occurs due to the fact that several links are not permanently used in

---

[1]The physical partitioning can be obtained since there is a relationship between each node in the physical system and each node in the information-sharing network as it has been presented in Remark 8.3 at Sect. 8.1.3

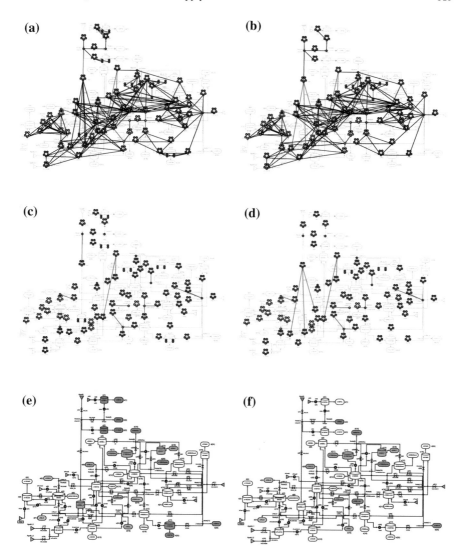

Fig. 9.6 Partitions for different time instants for the BWSN. Figures (a) and (b) correspond to optimal system partitioning for the information-sharing graph, Figures (c) and (d) correspond to the graph of links connecting partitions, and Figures (e) and (f) correspond to the respective physical partitioning. Moreover, Figures (a), (c) and (e) correspond to time instant $k = 15$, and Figures (b), (d) and (f) correspond to time instants $k \in \{40 - 42, 44 - 47, 64 - 66, 68 - 71, 88 - 90, 92 - 95, 112 - 114, 116 - 119\}$

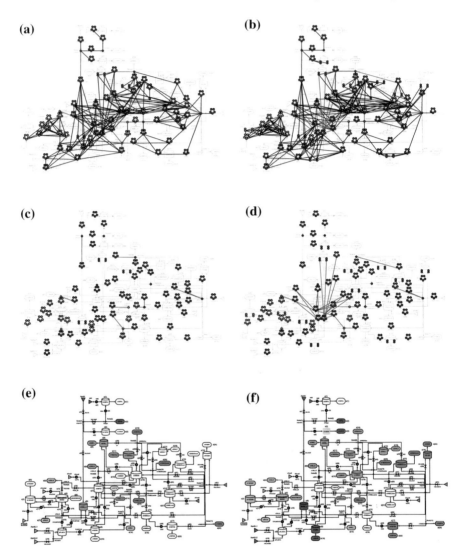

Fig. 9.7 Partitions for different time instants for the BWSN. Figures (a) and (b) correspond to optimal system partitioning for the information-sharing graph, Figures (c) and (d) correspond to the graph of links connecting partitions, and Figures (e) and (f) correspond to the respective physical partitioning. Moreover, Figures (a), (c) and (e) correspond to time instants $k \in \{43, 67, 91, 115\}$, and Figures (b), (d) and (f) correspond to time instants $k \in \{55 - 47, 79 - 81, 103 - 105\}$

the computation of the optimal control input at each time instant, i.e., links denoted by $\tilde{\mathcal{E}}$. Therefore, there is a compromise between the subtle variations to achieve the reference and the amount of permanent information-sharing links the controller uses. In this regard, when contemplating economical costs associated to the information-sharing links, it is convenient to implement control strategies with dynamical system partitioning. Finally, Fig. 9.5 shows the evolution of the information-sharing graph along the time, achieving a periodic behavior from the third day. It can be seen that the periodicity in the number of connected links is related to the periodicity of the demands (disturbances) presented in Fig. 6.8.

9.4   Discussion and Summary

Table 9.1 shows the KPIs corresponding to the costs associated to each actuator, and to the required communication links. It can be seen that the lowest economical costs $KPI_{costs} = 96.9812$ are obtained with the Scenario 1 corresponding to fixed constraints, i.e., with constant information-sharing network and constant static system partitioning. However, notice that the Scenario 1 also corresponds to the highest costs associated to the communication links $KPI_{links} = 73680$. In contrast, Scenario 2 is the one with lowest communication costs, i.e., $KPI_{links} = 49366$. Moreover, the Scenario 1 has lower economical costs $KPI_{costs} = 109.9567$ in comparison to $KPI_{costs} = 115.0512$ of the Scenario 2. In conclusion, notice that Scenario 2 is the best control strategy considering the overall performance shown in Table 9.1 if equal relevance is assigned to both KPIs or if more relevance is assigned to the communication links.

A general methodology to generate distributed density-dependent population dynamics has been presented by considering a reproduction rate in the distributed mean dynamics. Furthermore, it has been shown the relationship between the

Table 9.1 Summary of KPIs corresponding to operation of actuators and communication links

Day	Scenario 1		Scenario 2		Scenario 3	
	KPI_{costs} (e.u.)	KPI_{links} (e.u.)	KPI_{costs} (e.u.)	KPI_{links} (e.u.)	KPI_{costs} (e.u.)	KPI_{links} (e.u.)
1	22.6773	14723	24.2953	10797	24.2456	14736
2	19.6910	14723	22.2863	9666	21.6782	14736
3	18.2795	14723	22.9248	9671	21.4634	14736
4	17.8871	14723	22.5787	9565	21.3294	14736
5	18.4463	14723	22.9660	9667	21.2401	14736
Total	96.9812	73615	115.0512	49366	109.9567	73680
Overall*	73711.98		49481.05		73789.65	

*The overall cost is computed by adding the total costs for both KPIs, i.e., the total values for $KPI_{Ecosts} + KPI_{Ccost}$

equilibrium point of DDPG with the optimal point in a constrained optimization problem by selecting the description of benefits throughout the strategies using the Lagrangian of a potential function. In addition, the asymptotic stability of the equilibrium point under the D3RD, the D3SD, and the D3PD, has been formally proven for constant and time-varying population-interaction structures. Then, after introducing this class of dynamics and their properties, they have been applied to the design of a DMPC controller under a time-varying information-sharing network. On the other hand, a multi-objective partitioning procedure considering several aspects such as amount of links connecting different partitions, size of partitions, distance among elements, and importance of links has been presented in order to determine the appropriate partitioning in a large-scale system. As one of the most relevant features of the proposed partitioning is that it can be performed in a distributed manner. Therefore, the DMPC controller based on DDPG is combined with the distributed partitioning algorithm in two different manners, i.e., with static and dynamical system partitioning. The results for these two DMPC controllers based on DDPG and performing both static and dynamical system partitioning are presented, showing the effectiveness of both the DDPG approach and the partitioning for large-scale systems. As further work, the proposed non-centralized control design with partitioning can be tested in presence of faults at some partitions, so that the strategy facilitates the appropriate isolation.

References

1. Barreiro-Gomez J, Quijano N, Ocampo-Martinez C (2016) Distributed MPC with time-varying communication network: A density-dependent population games approach. In: Proceedings of the 55th IEEE conference on decision and control (CDC). Las Vegas, USA, pp 6068–6073
2. Barreiro-Gomez J, Ocampo-Martinez C, Quijano N (2017) Partitioning for large-scale systems: A sequential dmpc design. In: Proceedings of the 20th IFAC world congress. Toulouse, France, pp 8838–8843
3. Gupta A (1997) Fast and effective algorithms for graph partitioning and sparse-matrix ordering. IBM J Res Dev 41(1):171–183
4. Ocampo-Martinez C, Bovo S, Puig V (2011) Partitioning approach oriented to the decentralised predictive control of large-scale systems. J Process Control 21(2011):775–786
5. Barreiro-Gomez J, Obando G, Quijano N (2017) Distributed population dynamics: optimization and control applications. IEEE Trans Syst Man Cybern Syst 47(2):304–314
6. Quijano N, Ocampo-Martinez C, Barreiro-Gomez J, Obando G, Pantoja A, Mojica-Nava E (2017) The role of population games and evolutionary dynamics in distributed control systems. IEEE Control Syst 37(1):70–97
7. Barreiro-Gomez J, Quijano N, Ocampo-Martinez C (2016) Constrained distributed optimization: a population dynamics approach. Automatica 69:101–116
8. Chong EKP, Zak SH (2013) An introduction to optimization. Wiley series in discrete mathematics and optimization. Wiley, New York
9. Barreiro-Gomez J, Ocampo-Martinez C, Quijano N (2017) Dynamical tuning for mpc using population games: a water supply network application. ISA Trans 69(2017):175–186
10. Barreiro-Gomez J, Ocampo-Martinez C, Quijano N (2015) Evolutionary-game-based dynamical tuning for multi-objective model predictive control. In: Olaru S, Grancharova A, Lobo Pereira F (eds) Developments in model-based optimization and control. Springer, Berlin, pp 115–138

Part IV
Concluding Remarks

Chapter 10
Contributions and Concluding Remarks

10.1 Contributions

In this thesis, the role of some game-theoretical approaches for the design of optimization-based controllers (mainly DMPC controllers) has been presented. It has been shown that the classical population dynamics are appropriate to solve a resource allocation problem under the framework of full-potential games in Chap. 2. As a *first contribution* of this dissertation, and taking advantage of this, an on-line and dynamical tuning methodology based on population dynamics has been presented in Chap. 3.

In contrast, the population-games approach is unsuitable for the design of controllers that should consider more requirements associated to communication limitations and constraints. The classical population dynamics that require full information and that can deal with a unique coupled constraint and positiveness of variables have been studied. As a *second contribution* of this dissertation, the classical population-games approach is extended to the case in which strategy-constrained interactions are considered in Chap. 4 where a DMPC controller is designed, and a distributed engineering application, which illustrates the advantages of the contributions in Chap. 4, is presented in Chap. 5. As a *third contribution* of this dissertation, the classical population-games approach is extended to the case in which more coupled constraints can be considered, incorporating birth and death rates within the population games in Chap. 6. Therefore, these contributions regarding population games and optimization have permitted to consider the distributed population dynamics as a powerful tool in the design of DMPC controllers.

Regarding the cooperative/coalitional games, the power index known as the Shapley value is studied in Chap. 7. The main issue that has been addressed in this thesis is the high computational burden caused by the combinatory explosion of the traditional operations to find the Shapley value. To this end, the *fourth contribution* of this dissertation consists in considering a particular characteristic function, which is appropriate and commonly used for control purposes, allowing to reduce the computational costs to compute the power index from the order of hours to seconds. Also,

J. Barreiro-Gomez, *The Role of Population Games in the Design of Optimization-Based Controllers*, Springer Theses, https://doi.org/10.1007/978-3-319-92204-1_10

it is shown that the novel proposed approach can be performed under a distributed communication structure.

Over the last part of this thesis, it is discussed the partitioning of large-scale systems as a useful methodology in the design of non-centralized controllers. As the *fifth contribution* of this dissertation, it is proposed to perform the partitioning algorithm taking into account the information-sharing network that is required to compute the control inputs for a specific control strategy in Chap. 8. In this regard, the partitioning can be performed over any system and for any control technique, being a general approach. Furthermore, it is shown in Chap. 9 that it is possible to combine the contributions presented in Chaps. 6 and 8 to develop a dynamical system partitioning for the design of DMPC based on population games.

10.2 Answering the Research Questions

The conclusions are synthesized by answering the *key research questions* presented in Chap. 1 as follows:

(Q1) *Which kind of constrained optimization-based controllers can be designed by using the classical population dynamics and what are the information requirements?*

It has been shown in Chap. 2 that the classical population-games approach can be used for the design of resource-allocation controllers. This is made by exploiting the relationship of the Nash equilibrium with the optimal point in a constrained optimization problem (2.13) of a stable and full-potential game (see Theorem 2.1). Regarding the information requirements, the evolution of the proportion of agents in the classical population dynamics requires information about the fitness functions for the whole population (e.g., see the sum term in (2.14), (2.15), (2.16), and (2.17)). In this regard, when solving a resource-allocation problem with this approach, it is necessary to guarantee that all the sub-systems receiving the limited resource can share information to one another. In other words, the classical population dynamics represent a centralized controller whereas the distributed population dynamics presented in Chap. 4 represent a distributed controller.

(Q2) *How to develop a dynamical tuning methodology for MPC controllers with low computational burden?*

One of the main inconvenients of some on-line tuning methodologies for MPC controllers is that they require to find multiple points over the Pareto front implying high computational burden. Chapter 3 has presented a population-games based dynamical tuning for the prioritization weight parameters of a multi-objective MPC controller. The proposed methodology is performed with low computational burden since it only requires to know one point in the Pareto front corresponding to the current value of the control objectives in the cost function.

(Q3) *How to reduce the amount of required information in the evolution of population dynamics?*

The classical population dynamics are deduced from the *mean dynamics* by considering that each agent that receives a revision opportunity can select another agent who is playing any other strategy, i.e., when taking a sample of agents from the population, then the probability that the sampled population contains each strategy is the same. This thesis has shown that it is possible to consider strategy-constrained interactions, i.e., when an agent receives a revision opportunity, this agent can only select an opponent from a certain subset of agents playing specific strategies. In this regard, it has been shown that, from the *mean dynamics with strategy-constrained interactions* and by using different revision protocols, the distributed population dynamics are deduced as presented in Chap. 4. Therefore, the information requirements are reduced and preserving the relevant population-dynamics properties, i.e., the invariance of the simplex set and the stability of the equilibrium point.

(Q4) *How can the population-games approach be used in the design of distributed optimiza-tion-based controllers?*

The main feature exploited in this thesis in order to design optimization-based controllers by using a population-games approach is that, under the framework of full-potential games, the Nash equilibrium corresponds to a extreme point of the potential function as presented in Chap. 2. In this regard, it is possible to solve the constrained optimization problem in (2.13) by using this game-theoretical approach. Moreover, once the information requirements have been relaxed according to the answer to *key research question* (Q3), then the constrained optimization problem can be solved in a distributed manner. Furthermore, the optimization problem presented in (2.13) coincides to the form of a resource-allocation problem, i.e., this tool is appropriate to control systems in which there is a limited resource as presented in Chap. 4, where a DMPC controller is designed. On the other hand, Chap. 5 shows a control application based on population games, i.e., a distributed formation control under time-varying communication network is designed in a leader-follower fashion and by taking advantage of the properties of the distributed population dynamics.

(Q5) *How can more coupled constraints be considered with the population-games approach in order to make them suitable for a larger variety of problems in the design of distributed optimization-based controllers?*

The classical and distributed population dynamics are deduced from the *mean dynamics* and the *distributed mean dynamics*, respectively. One of the features of the mean dynamics is the mass conversion, i.e., any proportion that adopts a new strategy is leaving another strategy maintaining the population mass constant. This fact makes population dynamics satisfy the invariance of the simplex (2.9), and therefore satisfy the constraint (2.13b) along the time. Nevertheless, the mass conservation does not allow to consider more coupled constraints different from (2.13b). On the other hand, under the framework of density games, the population mass evolves along the time. To this end, it is proposed to consider a reproduction rate in the *mean dynamics*, i.e., a version of the *distributed density-dependent mean dynamics* is introduced in Chap. 6 permitting the deduction of the *distributed density-dependent population dynamics*. The density-dependent population dynamics also evolve to a Nash equilibrium; however, this equilibrium is achieved with a population mass such that the reproductions rates are null, i.e., a new condition is incorporated. Therefore, Chap. 6 has shown that these new properties can be conveniently used to solve constrained optimization problems with multiple coupled constraints enlarging the spectrum of types of systems that can be controlled by this game-theoretical approach, e.g., distributed-density games can be used in the design of DMPC controllers as presented in Chap. 6.

(Q6) *How can the computational burden associated to the computation of the Shapley power index be reduced and how it can be found under a distributed structure?*

The Shapley value is one of the most used power indices in the context of cooperative games, which is computed by defining a characteristic function and by considering the contribution of each player to all the possible coalitions that can be formed. This power index has been widely used in the design of controllers since it can determine the importance and/or influence of subsystems, communication elements, among others. However, the main issue when computing the Shapley value is the high computational burden due to the combinatorial explosion specially in the case of several players, e.g., for large-scale system applications. In order to reduce the computational burden of the Shapley value computation, it has been defined a specific characteristic function that can be associated to an error in the control context. By restricting the characteristic function, it is possible to find an alternative linear calculus that is equivalent to the conventional computation of the Shapley value, which requires low computational burden as it is formally presented in Chap. 7. Therefore, the computational time is reduced from hours to seconds. Moreover, due to the fact that the Shapley value computation results simpler, then it is possible to compute it under a distributed structure by using consensus ideas to propagate information. This result permits the use of the Shapley value in the control context for large-scale systems and with non-centralized control configurations.

(Q7) *How can the partitioning of large-scale systems be performed in a distributed manner and how it helps in the design of decentralized optimization-based controllers?*

Most of the partitioning methods use a graph representation of systems, which might make these methods quite specific for some types of systems, e.g., flow-based systems. In contrast, it has been proposed to perform a partitioning algorithm based on the information-sharing graph making the methodology general for any system and control strategy. Furthermore, due to the fact that the information-sharing graph also illustrates how the dependence among decision variables is in order to design a non-centralized controller, it is proposed a distributed partitioning algorithm. To this end, it has been shown in Chap. 8 that the required operations in the algorithm can be computed in a distributed manner. In addition, since each node in the information-sharing graph can make the decision to select a partition, then the computational burden is reduced.

(Q8) *How can the partitioning of a large-scale system be performed dynamically and how can the population-games approach be used in the design of partitioned optimization-based distributed controllers?*

The appropriate partitioning of a large-scale system might depend on the current system state, or it might be also affected by disturbances. Therefore, it might be convenient to implement a dynamical partitioning strategy. According to the answer for the *key research question* (Q7), the computational burden for the proposed partitioning algorithm is low and the information requirements to perform the partitioning can be executed in a distributed manner. Therefore, it is an appropriate tool to make a dynamical partitioning in large-scale systems. In this regard, it is necessary to have a control strategy that can deal with time-varying information-sharing graphs. Then, it is plausible to combine the distributed control design presented in Chap. 6 and the partitioning algorithm introduced in Chap. 8. Therefore, a DMPC based on population games and performing a dynamical system partitioning is presented in Chap. 9.

Biography

Julian Barreiro Gomez was born in Colombia in 1988. He received his B.S. degree (cum laude) in Electronics Engineering from Universidad Santo Tomas (USTA), Bogota, Colombia, in 2011; the M.Sc. degree in Electrical Engineering and the Ph.D. degree in Engineering from Universidad de Los Andes (UAndes), Bogota, Colombia, in 2013 and 2017, respectively; and the Ph.D. degree (excellent cum laude) in Automatic, Robotics and Computer Vision from the Technical University of Catalonia (UPC), Barcelona, Spain, in 2017. He is currently a Post-Doctoral Associate in the Learning and Game Theory Laboratory (L&G lab), Engineering Division at the New York University in Abu Dhabi (NYUAD), United Arab Emirates. His main research interests are Mean-field-type Games, Differential Games, Stochastic Games, Constrained Population Games and Evolutionary Dynamics, Partitioning of Large-scale Systems, Distributed Optimization, and Distributed Predictive Control

© Springer International Publishing AG, part of Springer Nature 2019
J. Barreiro-Gomez, *The Role of Population Games in the Design of Optimization-Based Controllers*, Springer Theses,
https://doi.org/10.1007/978-3-319-92204-1

Printed in the United States
By Bookmasters